Nuclear power and public responsibility

Nuclear power and public responsibility

L.E.J. Roberts

Director, Atomic Energy Research Establishment, Harwell, England

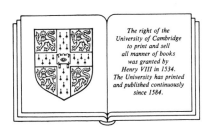

The right of the
University of Cambridge
to print and sell
all manner of books
was granted by
Henry VIII in 1534.
The University has printed
and published continuously
since 1584.

CAMBRIDGE UNIVERSITY PRESS

Cambridge

London   New York   New Rochelle

Melbourne   Sydney

Published by the Press Syndicate of the University of Cambridge
The Pitt Building, Trumpington Street, Cambridge CB2 1RP
32 East 57th Street, New York, NY 10022, USA
296 Beaconsfield Parade, Middle Park, Melbourne 3206, Australia

© Cambridge University Press 1984

First published 1984

Printed in Great Britain at the University Press, Cambridge

Library of Congress catalogue card number: 84–7644

British Library cataloguing in publication data
Roberts, L.E.J.
Nuclear power and public responsibility.
1. Atomic power-plants – Great Britain –
Safety measures
I. Title
363.1′79   TK1362.G7
ISBN 0 521 24718 7

Contents

Preface

The evolution of safety standards and criteria in the industry that supports the commercial generation of electricity from nuclear power stations is a continuing process that rests on a broad basis of scientific investigation and risk assessment. Three main themes can be identified as major aims in this field of responsibility: the need to protect workers and the public against the effects of toxic emissions and also against the consequences of accidents, and the need to protect the environment against irreversible change. These themes are in fact applicable to many other industrial contexts besides nuclear power. Technologists are now expected – more strictly than ever before – to minimise any harmful effects of the developments they wish to pursue. It is increasingly important in the modern age to be able to strike a balance between the benefits foreseen from new technical developments and the resources that have to be deployed to achieve or to prove an acceptable level of safety.

This book is a revised and slightly expanded version of the 1981 R.M. Jones' Lectures at Queen's University, Belfast. The lecturer is asked to illustrate some aspect of the evolution of ideas. The University invites a number of distinguished guests and many members of the University to a discussion following each lecture, and I am most grateful to those who attended and contributed much to the eventual material for this book. Above all, my thanks are due to the Vice-Chancellor, Dr Peter Froggatt, and to Mrs Froggatt, for their kindness and hospitality which made my stay in Belfast so very pleasant.

This is a book for the lay reader, not for the expert. Some references are included to enable the interested reader to look further into the subject but no attempt has been made to provide a comprehensive bibliography; there is a very wide and extensive technical literature. I am indebted to many of my colleagues in Harwell and elsewhere in the Atomic Energy Authority for a great deal of help and constructive criticism, and in particular to Mr P.A.H. Saunders for the contributions that he made and for editing the entire text and assembling the illustrations.

I hope the facts stated are correct. The opinions voiced are entirely my own and are not to be construed as 'an official view'.

L.E.J. Roberts
August 1983

1
The scientific and historical background

Some historical perspectives

Concern about the effects of new technologies and of new products is no new thing. 'An intolerable smell diffuses itself throughout the neighbouring places and the air is greatly infected to the annoyance of the citizens and to the injury of their bodily health', read a proclamation to the lime burners of Southwark in the year 1307; the subject was the burning of coal – and the proclamation preceded the Clean Air Act by some six centuries. Similar forebodings about the introduction of petroleum were voiced in the Congress of the US in the 1880s. Industrialisation based on the exploitation of these fossil fuels has led to great economic gains, with the improvements in health, mobility, literacy and the general quality of life with which we are all familiar. With the benefit of hindsight, most would conclude that these enormous general benefits have outweighed the risks and the ravages of the consequent pollution, serious though those were and still are. Along with a rising standard of living has come a flood of new products of every kind, which we all enjoy, though many would conclude, again with hindsight, that King James I of England and VI of Scotland was right to express his horror of one product new in his day, namely, tobacco.

The economic advances of industrialisation have been gained at a considerable cost in human lives and misery. The conditions in industry at the beginning of the industrial revolution were harsh indeed, and were tolerated only because of sheer necessity and because the conditions of the rural labourer at the time were as bad or worse. The idea that people engaged in industry have a responsibility to provide safe working conditions for their workers grew slowly; conditions were gradually ameliorated during the 19th century as a result of public pressures and the action of enlightened employers. Legislation was introduced, first to protect young people and then more generally, and the first medical inspector of factories was appointed in 1898. The development of a sense of responsibility for the general environment came rather later. Action during the 19th century was limited to smoke abatement in some urban areas and the first major piece of legislation was the Alkali Act of 1906.

During this century we have seen an unprecedented rate of innovation, caused by the conscious application of science to technological ends, and by the rapid introduction of new processes, products and technologies. The rate of change has been particularly marked during the last four decades. Scientific advances have also enabled more precise measurement to be made of the environmental and human consequences of the application of technologies, both old and new, and this has in turn resulted in deeper questioning of these consequences. Ideas on the control of technology, on the responsibilities of technologists and on their public accountability, have evolved rapidly and there has been a parallel institutional evolution through which these ideas may be expressed in regulations. The extent of the economic use of many technological developments – insecticides, fertilisers, drugs, plastics, air travel, nuclear power – is now largely determined by the measures required for the protection of the public and of the environment. It is therefore vital that a proper balance and perspective be achieved between gain and detriment or much misapplication of effort and loss of potential benefit can ensue.

In no industry have these issues been more vigorously

debated than in the commercial development of the generation of electricity by nuclear power. Many aspects of this development have caused acute and polarised controversy in their own right, but the underlying themes of regulation based on sound technical knowledge, of public accountability and of a proper degree of governmental control are common to many technologies. The evolution of the concepts of regulation in the nuclear industry is therefore a subject of more general interest; the same fundamental questions are being raised in many other industrial contexts, as we shall see.

Opposition to nuclear power is not always based on mistrust of the technical basis of safety arguments and of the regulations made for the protection of the public. For instance, the nuclear industry is seen as representative of large, over-centralised, capital-intensive technologies by some who are opposed to, or worried by, these trends in modern industrialised societies for a variety of reasons. Still others fear that the development of the civil uses of nuclear power will lead to the spread of nuclear weapons and increase the risk of the use of nuclear bombs in war. Considerations such as these go well beyond the responsibilities to their own workers, and to the public, which can reasonably be laid upon the scientists and technologists serving the civil nuclear power industry. For example, the economies that can be achieved by increasing the scale of production plants dictates the concentration of many industrial processes into large units in the modern world, and the generation of electricity from nuclear or fossil fuels is no exception. Society could be served by smaller units if increased costs were acceptable, and this type of argument has little to do with nuclear power as such, but rather with large questions of the organisation and philosophy of an industrial society. Then again, the whole question of nuclear weapons is a matter for

international political control, as has been well realised since
the passage of the Non-Proliferation Treaty in 1970 and the
organisation of safeguards under the auspices of the International Atomic Energy Agency. The development of a civil
nuclear power industry is certainly not a pre-requisite to a
military programme; the five nations known to be equipped
with nuclear weapons – the US, USSR, UK, France and
China – all acquired the weapons before building electricity
generating stations and not as a result of having done so.

These various strands of argument must be separated if
the debate about nuclear power is not to become confused.
This book is concerned with the responsibilities for ensuring
public safety which are carried by technologists in the civil
nuclear power industry; with the evolution of the scientific
knowledge on which protective action must be based; and
with the necessary and parallel development of regulation
and of regulatory bodies. Three main themes are followed.
The first is the need to protect the public against dangerous
emissions of toxic substances. The nuclear industry is concerned with protection against radiation and radioactive substances, but the type of problems encountered in the definition of safe levels of radiation are identical to those met in
the chemical industry generally and in medicine, when one
has to define an acceptable level of exposure to an agent
known to be toxic at high concentrations. The second theme
is the protection of the public against the consequences of
accidents, the growth of the subject of risk analysis as a
branch of engineering, and the parallel study of the perceptions of risks by the lay public. The third main theme is the
responsibility of technologists towards the environment in
general, and the consequent need to dispose of waste products safely and in a way which will not constitute a threat to
future generations.

Over the last 20 years, there has been notable growth

both in the concern for personal safety and of the feeling that human beings have been entrusted with the stewardship of 'spaceship Earth'. Such an increase in personal and collective responsibility is wholly welcome, but there is a price to be paid. No one can be promised immortality and every human activity carries some degree of risk. The nuclear industry already spends a great deal, relatively speaking, to attain a very high degree of safety. The discussion of the major themes mentioned above is therefore not complete without some attempt to relate the costs of additional safety measures to the estimated or perceived benefit, a theme which is taken up later. In this chapter, we shall look at the state of nuclear technology, and trace briefly the scientific history of the 80 years or so of radiation science.

Nuclear power as an energy resource

There are some 293 nuclear reactors now working, producing electricity in 24 countries, and a further 219 stations are being built (Table 1.1). These reactors will have a total

Table 1.1. *Nuclear power reactors in operation and under construction at end 1982*

Area	In operation		Under construction	
	Number of units	Capacity GW(e)	Number of units	Capacity GW(e)
OECD	231	144.5	145	145
Centrally planned economies	53	22.5	54	41
Other	9	3	20	14
Totals	293	170	219	200

Source: IAEA

capacity of about 370 GW(e);† they should all be completed
in this decade. By that time nuclear stations will produce
about 25% of the electricity production in OECD countries.

The reasons behind the drive to develop this technology, despite the confused, acrimonious, and sometimes violent debate which has arisen, are all related to the realisation
that the world's economic system has become overwhelm-
ingly dependent on the fuels – oil and gas – which are in the
shortest supply and most vulnerable to depletion. At present,
50% of the world's primary energy comes from oil. On the
basis of predictions made at the last World Energy Conference,[1] this proportion is expected to reduce to 30% by the
year 2000 and to 20% by 2020. Other energy sources will
therefore be needed in increasing amounts in the coming
decades and nuclear energy is one such source, already a
developed commercial option, able to compete with others.

The extent of the future contribution from any one
energy source cannot be predicted; it depends on population
growth, on the course of the world economy, on technological
developments in the supply and utilisation of fuels and on
national and international political factors which determine
whether resources are actually available for use. But in
simple terms there is no doubt that the world needs energy
and will need it in increasing quantities. This conclusion
seems inescapable if the standards of life in the poorer parts
of the world are to improve; Table 1.2 illustrates the very
unequal pattern of energy use across the world and the close
correlation between energy use, economic wealth and other
indicators of standard of living. At present, one-third of the
world population consumes two-thirds of the world's energy
supply and so to bring everyone up to the current average

†1 Gigawatt (electrical) – GW(e) – equals 1000 megawatts. A 1 GW(e)
  station will meet the electricity needs of a typical English county –
  about one million people.

Table 1.2. *Energy consumption in the world*[2, 3]

	Population (1975: millions)	Gross national product per capita ($)	Annual energy consumption per capita (kg coal equiv.)	Life expectancy at birth (years)	Infant mortality (deaths per 1000 live births)
Low income group (45 countries)	1130	166	90	48	135
Low–middle income group (38 countries)	1230	430	480	62	76
Upper–middle income group (38 countries)	505	1215	1110	62	85
High income group (48 countries)	1010	4980	4970	71	20

standards of the developed world would require a doubling of supplies. If we allow for expected population growth, then about six times the present energy consumption would be needed in 2025 to achieve the same per capita target, because most of the population growth is expected to occur in the groups now using the least energy. Such a target might be beyond our reach on this timescale, and it could be and should be reduced by determined conservation efforts and by higher efficiency in energy use. However, a large demand will undoubtedly remain, particularly if one includes an allowance for any growth in the per capita demand in the developing countries.

The prediction of the reserves of fuel which might satisfy this demand is notoriously difficult, since the important question is not what may exist in various parts of the earth's crust but how much can be obtained at economical prices, and that depends on the efficiency of the means of extraction and on other factors determining availability. Figure 1.1 shows some estimates of 'reasonably recoverable' reserves. Despite the uncertainties, two conclusions are clear. The first is the relatively poor position of the resources of our hydrocarbon fuels, oil and gas, which at present supply more than one-half the world's primary energy needs. The second is the potential importance of the uranium reserves, which is critically dependent upon the technology we use to exploit them. The energy equivalent of the known uranium reserves amounts to about one-half of the total estimated oil reserves if the uranium is used only in 'thermal' reactors of the types being built on a large scale today; it thus constitutes a very valuable, but not an enormous, additional resource. 'Fast' reactors† would transform the situation by increasing

---

†The descriptions 'thermal' and 'fast' relate to the speed of the flux of neutrons in the reactors. The basic factors of reactor design are discussed later in the chapter and the types of reactor being built or projected in the UK are illustrated in the Annex (see Figures A1.1, A1.2 and A1.3).

Fig. 1.1. World reserves of coal, oil, gas and uranium.

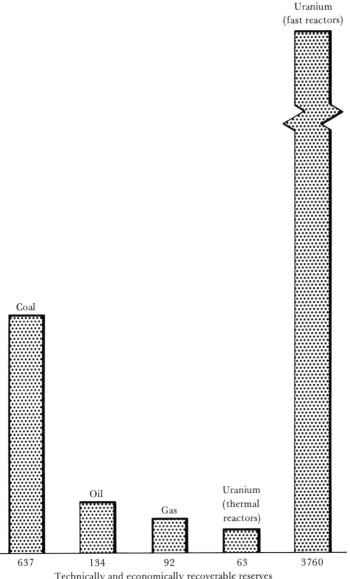

the energy recoverable from the same uranium reserves by a factor of 50 or 60, to the equivalent of over 20 times the oil reserves. The key difference is that thermal reactors can only utilise a small fraction of the uranium in the fuel whereas fast reactors eventually utilise a very much larger fraction, using material which is simply a waste product of the fuel cycle used in thermal reactors. For example, the quantity of such 'waste' uranium that we already have in the UK would, if used in fast reactors, provide an amount of energy comparable with that obtainable from our entire coal reserves.

In its present state of development, nuclear power is best fitted to the generation of base load electricity in large power stations. In the medium term, therefore, it is not a substitute for all the energy needs supplied by oil and gas, though it is for some of them. The use of nuclear power in those economies that can absorb large amounts of electricity for diverse purposes will reduce the pressure on the hydrocarbon reserves by the industrialised countries, and may be seen as an essential component of a transition to a mixed energy economy in which our fossil fuel reserves are increasingly reserved for those uses for which we have no ready alternative, such as transport and the production of petrochemicals; there is no other large-scale need for uranium. How far nuclear power is used in this way will depend on the relative economics of electricity generation from different fuels. This book is not concerned with the cost of nuclear power as such; it is sufficient to note that nuclear power is perceived in many countries as being commercially competitive or as offering considerable cost advantages in the medium or longer term.

The science of nuclear power

The development of weapons of unparalleled power and of an entirely new source of energy for peaceful purposes from the discovery of a new type of nuclear reaction, the 'fission' reaction, by two radiochemists (Hahn and Meitner), is a story of great drama and human interest, but it has been told many times, and it will suffice to give a very brief summary here. The essential discovery was that some isotopes of the heaviest elements undergo a particular type of nuclear reaction under bombardment with neutrons, the 'fission' reaction, in which the nucleus of the atom splits into two more-or-less equal parts, with the emission of a great deal of energy and more neutrons. The amount of energy released is enormously greater than that released in chemical reactions. If the geometrical arrangement of material is such that these neutrons can in turn effect the fission of more nuclei, then a chain reaction occurs with a very large release of energy. This release can take place with explosive force – an atomic bomb – if several pieces of pure or nearly pure fissile material are forced together to constitute a mass of greater than a critical size from which the neutrons cannot escape and therefore are multiplied rapidly by the chain reaction.

A controlled release of energy occurs in a nuclear reactor by assembling fissile isotopes in a dilute form in such a way that the number of neutrons responsible for the chain reaction can be controlled by inserting material that can capture neutrons without undergoing fission, like cadmium or boron. Because the fissile material in a nuclear reactor is used in a dilute form, a reactor can never explode like an atomic bomb. The only fissile isotope that occurs to any significant extent in nature is U^{235}, which constitutes only 0.7% of natural uranium, the rest consisting of the heavier isotope U^{238}.† The neutrons that result when U^{235} undergoes fission

†Naturally occurring chemical elements consist of a mixture of 'isotopes', that is, of atoms of the same chemical species but different atomic weights. 'U^{235}' denotes the isotope of uranium of atomic weight 235 units and similarly for 'U^{238}' and 'Pu^{239}'.

can suffer three fates: some must be used to cause more fissions and maintain the chain reaction; some escape or are captured by other materials; and some are captured by U^{238}, eventually forming another fissile isotope, Pu^{239}, an isotope of the new element, plutonium, which can be used as fissile material in place of U^{235}.

The nuclear reactors which are used today for electricity generation are 'thermal' reactors in which the neutrons responsible for keeping the reactor going are slowed down to low or 'thermal' energies by passing through material like carbon or water, which slows down the neutrons without capturing them. The 'Magnox' reactors built in the UK and the Canadian 'CANDU' reactors use natural uranium as a fuel. Reactors such as the pressurised water reactor (PWR) or our advanced gas-cooled reactors use uranium fuel which has been artificially enriched in U^{235} to about 3%. All these 'thermal' reactors are limited to the use of about 1% of the original uranium. A much higher efficiency of uranium use can only be achieved by converting the U^{238} isotope into the fissile material Pu^{239}. This can be efficiently achieved in 'fast' reactors – that is, reactors in which the fast neutrons are used to induce fission without the need to slow them down. The type of fast reactor which has been developed in the UK and other countries uses a mixed uranium–plutonium fuel consisting of about 20% plutonium, and the neutron economy of the reactor can be arranged either to use up the plutonium, or, by absorbing the escaping neutrons in U^{238}, to manufacture more plutonium from uranium as the nuclear reactions proceed. In this way almost all the uranium is used in time, rather than only 1% of it.

The fragments that result from the fission of U^{235} or Pu^{239}, the 'fission products', are radioactive and are the major source of radioactivity from nuclear power. Other radioactive materials arise from the capture of neutrons by other materials in the reactor core, and isotopes of plutonium and other heavy elements arise from the interaction of neutrons with uranium and plutonium. These radioactive materials are similar to the range of naturally radioactive isotopes that occur in nature, such as isotopes of potassium, thorium and radium. The neutrons in nuclear reactors are used to manufacture artificially radioactive materials for use in industry or medicine, such as cobalt or iodine, in radioactive forms.

There are, then, two major distinguishing characteristics of nuclear power which affect the external environment and which are inherent in the fission reaction itself – first, the great concentration of this energy source and, second, the accompanying production of high levels of radioactivity. The effects of the first, concentration, are almost entirely beneficial. The energy release from one ton of uranium used in thermal reactors is equivalent to that gained by burning 20 000 tons of coal. Since usable uranium ores contain 0.1% of uranium, or more, less than $\frac{1}{20}$ as much material has to be mined for the production of a given quantity of energy and the amount of waste that has to be transported and disposed of is much less, about $\frac{1}{200}$ of that in the case of coal, for the same energy production.

The high levels of radioactivity associated with the fission process, however, constitute the major potential impact of the industry on the environment and on the health of its workers and of the public. The safe management and regulation of the use of radioactive substances has been a main responsibility of the nuclear industry from the beginning, affecting the design of all nuclear installations, the management of workers, the economy of the whole industry and its impact on the public and on the environment.

The effects of ionising radiation

It was perhaps a fortunate accident that the emergence of an industry which necessarily involved the handling and control of large quantities of radioactive material took place more than 40 years after the nature of radioactivity had been recognised and after the use of radioactive substances and of ionising radiation had become widespread in small-scale industry and, most importantly, in medical applications, both for diagnosis and for treatment. It is commonly the case that an industrial hazard is recognised as a result of experience after fairly widespread use – the danger from asbestos is a case in point. But this stage had already been passed in the case of radioactivity some decades before the discovery of fission.

In fact, there was very little delay between the discovery of ionising radiations and the realisation that they could harm human tissue. X-rays were discovered by Roentgen in 1895, and the Curies reported the isolation of radium in 1898. Minck investigated the killing effects of X-rays on bacteria in 1896 and in the same year Lord Lister, addressing the annual meeting of the British Association, said this of X-rays: 'If the skin is long exposed to their action it becomes very much irritated, affected with a sort of aggravating sun-burning. This suggests that transmission through the human body may not be altogether a matter of indifference to the internal organs.'[5] In 1897 a US court awarded damages to a patient affected by over-exposure to X-rays. Rutherford observed when he visited the Curies in 1903: 'We could not help but observe that the hands of Professor Curie were in a very inflamed and painful state due to exposure to radium rays.'[6] Marie Curie later died of leukaemia.

Radiation only causes damage if it is absorbed by tissue – we are all bombarded by neutrinos from the sun, but since they go straight through us without interacting they cause no damage. The quantity of radiation that is absorbed is called the dose – the unit is the rad, a measure of a standard amount of energy absorbed per unit weight of tissue. The different types of common emissions from radioactive elements have very different penetrating powers. Gamma rays are quanta of electromagnetic radiation similar to but usually more energetic than X-rays, and γ-emitters commonly need shielding by several feet of concrete or steel. Beta particles are electrons and can penetrate only thin metal sheets. Alpha particles (the charged nuclei of helium atoms) cannot penetrate thin paper, or human skin, so α-emitters are not dangerous to people unless they are taken into the body. Neutrons, being of zero charge, have considerable penetrating power. Not all these types of radiation are equally effective in causing damage. For example, α-particles deposit their energy more rapidly than β- and γ-particles; thus, for a given absorbed dose, one would expect the α-particles to cause more damage. It is convenient for purposes of radiation protection to define a unit of radiation dosage which would be expected to cause the same amount of damage, irrespective of the type of radiation. We thus use the 'dose equivalent', which is the product of absorbed dose and a quality factor which expresses the different effectiveness of the different types of radiation in producing the delayed effects described below. The unit of dose equivalent is the rem.† In what follows, the word 'dose' is used as shorthand for 'dose equivalent'.

It may also be helpful here to indicate some yardstick against which different doses can be compared, and the most obvious is the natural background of radiation in which mankind has evolved. This arises from the cosmic rays reaching the earth from space, from the natural radioactivity of the rocks, from radon and its decay products in the air, and from the radioactive isotopes (mainly potassium^{40}) in our own bodies. An average value of this natural background in the

---

†New units for dose and dose equivalent, the gray and the Sievert, are now coming into use. One gray equals 100 rad, one Sievert equals 100 rem.

UK is just under ⅕ rem (200 millirem) per year, and the average life-time dose is about 12 rem.

Acute effects of large doses

Exposure to large doses of external or internal radiation quickly produces a variety of characteristic effects such as skin burns, radiation sickness and loss of fertility. Very large doses delivered to the whole body in a brief period – doses of several hundred rad in a few minutes – result in widespread damage to the gut, blood cells and bone tissues, and generally lead to death within a few days or weeks. Similar doses are used in radiotherapy for the treatment of tumors, but in a controlled way so that the absorbed dose is localised as far as possible. These effects are generally characterised by a threshold below which no effects are observed. The relationship between the dose of radiation received and the subsequent biological effects, the dose–response curve, is thus not a straight line, as is shown in Figure 1.2. For example, there is zero risk of early death at doses below about 100 rad, rising to a 50% risk at about 300 rad and levelling off at 100% risk at about 1000 rad.

Fig. 1.2. Dose–response curve – acute effects.

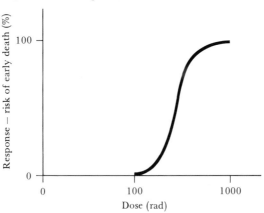

As the dose of radiation is reduced to below the level at which these acute effects are observed, the effects become rare and much more difficult to observe – they occur, if at all, only after delays of many years. The occurrence of these delayed effects must be studied by probabilistic techniques; they are therefore called 'stochastic' effects. Unlike the acute non-stochastic effects which were described in the last paragraph, they only occur in a small proportion of the irradiated subjects in an apparently random way. The size of the dose governs the probability of the effect occurring, but not the severity of the effect itself. The principal stochastic effects are the induction of malignancies in the exposed individual and the appearance of hereditary effects in later generations. There is no evidence of any life-shortening effects in individuals exposed to low radiation doses from any causes other than cancer. The delayed appearance of acute clinical effects in a small proportion of an exposed population is similar to the effects of other agents such as chemical carcinogens, asbestos or tobacco.

Stochastic effects – cancers. The evidence for a link between radiation exposure and human cancers comes from a number of particularly well-documented groups of people: the Hiroshima and Nagasaki bomb survivors, patients who have received large doses of therapeutic radiation, and workers who, in the past, have been exposed to high levels of radiation, such as radiographers, watch-dial painters using radium, and miners of uranium and some other ores working in inadequately ventilated mines and therefore exposed to high concentrations of radon. Much relevant information has also come from animal studies.

Studies of these exposed populations have shown that doses of the order of 100 rem result in a small increase in the 'natural' incidence of cancer. A major difficulty in arriving at precise estimates of the increase is the fact that cancer is by no means uncommon – in the UK it is the cause of one in five of all deaths. Moreover, cancer incidence is variable for reasons that are not understood; more people in urban areas die of cancer than in rural areas and there are considerable variations even between the populations of different towns. Some of the causes seem to be due to variations in environment and diet; for example, the incidence of stomach cancer is particularly high in Japan, but falls to the average US rate in Japanese immigrants to the US after the second generation.

The studies of human populations from which numerical estimates of the risk can be made are necessarily limited to those who have received radiation doses well above background levels. The total number of cancer cases that can be ascribed to radiation is rather small. The United Nations survey[7] reports that about 180 excess cases of lung cancer have been identified among some 15 000 uranium and other metalliferous miners; about 500 excess malignancies (not all fatal) have been ascribed to the radiation doses received by various groups of patients for different medical conditions, a total population of some 65 000 patients, mainly in the UK, US and Israel (Table 1.3). One of the most important studies is that of the Hiroshima and Nagasaki bomb survivors, where very detailed health records and estimates of actual doses are available. Here, 82 000 of the 285 000 survivors, about one-third, have been studied extensively for 35 years, the comparison group consisting of people who were in the cities at the time of the bombs but who were not exposed to large doses of radiation. Some 4000 of the 82 000 have now died of cancer, but comparisons with the control population shows that only about 200 of these cancer deaths ($1/20$ of them) can be ascribed to the radiation from the bombs. From a knowledge of the total dose received by the exposed population a figure for the number of cancer deaths per unit dose can be derived *if it is assumed that the effect is proportional to the dose*. The evidence from the bomb survivors and from the other groups is summarised in several major reports,[7, 8] and is also reviewed by the International Commission on Radiological Protection (ICRP).[9] The most recent ICRP estimate gives an average risk for fatal cancers of just over one case per 10 000 people all exposed to a dose of 1 rem, and from the range of other estimates it can be inferred that the uncertainty of this figure is a factor of two or three. The quantitative estimates are liable to be refined as new evidence accumulates;[10] the results from Hiroshima and Nagasaki have recently been subjected to a fresh examination, the results of which are not available yet in final form. What is important for the present argument is the type of evidence on which the conclusions are based.

We can conclude that enhanced cancer incidence has been observed as an infrequent, but undoubted, effect in groups exposed to doses of the order of 50 rem or above. There are a few special cases where excess cancers have been reported in groups exposed to doses of less than 10 rem but

Table 1.3. *Excess cancer deaths in some exposed populations*

Population studies		Excess cancer deaths
Occupational		
Radiologists	4000	10
Miners	14 000	180
Radium luminisers	800	50
Medical		
Diagnosis and therapy	50 000	400
Military		
Bomb survivors	80 000	200

Source: UNSCEAR[7]

no quantitative risk estimates can be made on the basis of these observations. The question, then, arises – what is the effect of smaller doses, typically less than 1 rem per annum, such as those encountered in practice in the nuclear industry and by medical workers with radiation and of the even smaller doses received by the public as a result of the operations of the nuclear industry?

The problems in finding answers to these questions are indeed formidable. Consider first the simple question of statistics. We are dealing with a phenomenon with an occurrence of about one in 10000 in a population irradiated to a level of 1 rem. To achieve reasonable statistical accuracy, populations of the order of 100000 or more have to be studied and achieving proper statistical comparison with a control population is itself a difficult problem. There is no one population that would be expected to yield an answer with high statistical certainty. However, other reasons may indicate that such studies should be attempted. The Atomic Energy Authority has recently commissioned the Medical Research Council to undertake a mortality study of its own employees and ex-employees; this is a population of 45000 in which the exposure records are better known than most. The results should be available in 1984. Larger studies such as the National Registry, established by the National Radiological Protection Board to include all radiation workers in the UK, will report in some years time.

In addition to these groups of people who have been exposed to man-made sources of radiation, it is possible to find national populations which have been exposed to more than the average radiation resulting from local geology or from the variations of cosmic ray intensity with altitude. Attempts have been made to find correlations between natural cancer incidence and background radiation levels, and no statistically significant effects have been found.

Indeed, in the US it is found[11] that states with a high level of background radiation, like Colorado, have a significantly lower number of cancer deaths than the eastern seaboard states where radiation levels are lower by about a factor of two, as illustrated in Figure 1.3. This does not mean of course that radiation is necessarily good for you, but simply confirms that these background levels are not a major cause of cancer. This negative result is not surprising; if the rates of cancer incidence deduced above can indeed be extrapolated to background doses, only about 22 people in a population of one million would die in any year as a result of background radiation, and a variation of this figure, even by a factor of two, would be statistically insignificant, given that about 200 000 in that population will eventually die of cancer for other reasons.

The further investigation of low-dose risks must therefore involve indirect methods through the development of fundamental knowledge in radiobiology and the knowledge of mechanisms of action from which dose–response models may be derived. While radiobiological studies in the laboratory clearly show the damage that radiation can cause to living cells, the body is in fact remarkably resistant to carcinogenesis due to radiation. One week of natural background radiation will result in one ionisation event – a radiation-induced reaction that is a potential cause of damage – in every one of the body cells. But the natural background cannot be the cause of more than a very small fraction of all cancers, as we have seen, and therefore either very nearly all of the damage in the cells must be successfully repaired, or it must almost all be innocuous and only extremely rarely contribute to the generation of a cancer. The fundamental reason for the appearance of cancer in a very few individuals in an irradiated population after a long latent period is not known; the same problem and the same

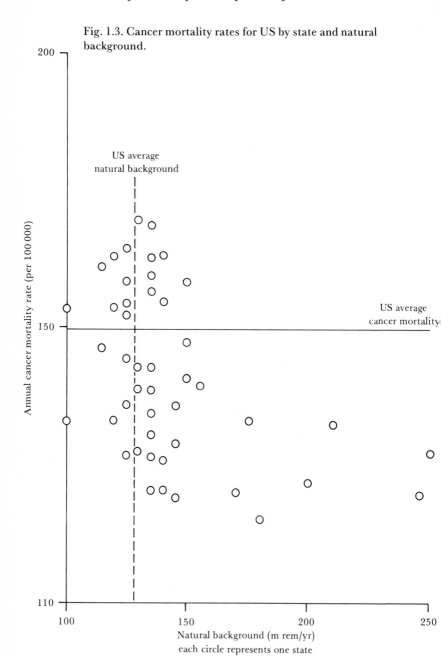

Fig. 1.3. Cancer mortality rates for US by state and natural background.

degree of ignorance exists also in the case of cancer due to chemical agents.

Faced with the need to extrapolate the available data to lower doses where effects are not, and probably can never be, directly observable, the common and cautious assumption has been that these delayed effects are linearly dependent on dose down to zero dose, with no threshold. This linear hypothesis is unlikely to be correct in general and several alternative hypotheses have been suggested. There is some experimental evidence for the theory of dual radiation action, which predicts a quadratic dose–response for some types of radiation – that is, that the number of cancer cases will increase as the square of the radiation dose. Some major studies have concluded that the linear hypothesis is more likely to lead to an over-estimate than an under-estimate of the risks of low doses of radiation, particularly for X-rays, γ-rays and electrons. The most recent expression of this view is given in the 1980 Report of the Committee on Biological Effects of Ionising Radiations (BEIR) instituted by the US National Academy of Sciences,[8] which concluded, albeit without unanimity, that a linear quadratic model, inter-mediate between a pure linear and a pure quadratic model, was the most consistent with the epidemiological and radio-biological data for γ- and X-rays. The different dose–response curves are illustrated in Figure 1.4; the linear extra-polation gives the most pessimistic values at low doses. The linear model was thought to be appropriate for α-particles and neutrons. These conclusions were supported by 14 out of the 16 members of the sub-committee on somatic effects and the final Report included two dissenting opinions. The first, by the Committee's Chairman, Edward Radford, is a defence of the use of a simple linear no-threshold model as appropri-ate at the present time, given the complexity of the problem and the great variability of the human and environmental

Fig. 1.4. Dose–response models – stochastic effects.

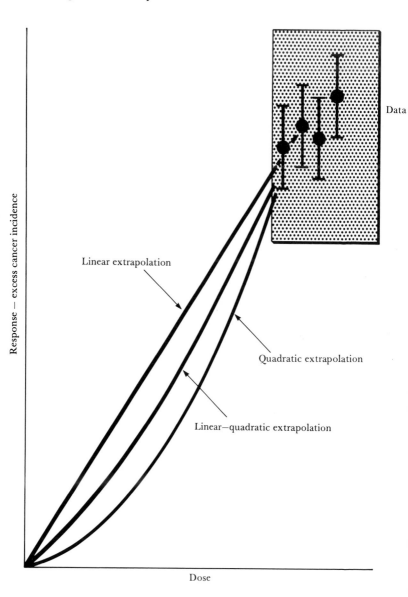

factors capable of influencing the results. He said: 'Until we know more about the process of cancer development in man we cannot go further with this problem.' In contrast, one member of the Committee, Harald Rossi, advances the argument that the most plausible risk from X-rays, γ-rays and electrons is lower than that given by the extrapolation method advocated in the main report and argues that a pure quadratic law is a better fit to the experimental data.

The conclusions of these many studies and the basis for extrapolation are not accepted by everyone and the evolution of this subject has been marked by quite violent controversies. The best known of these were summarised by the BEIR Committee[8] who published their own analysis and also their comments on several studies which appeared to conflict with their conclusions, such as the Mancuso, Stewart and Kneale study of cancer mortality at the Hanford Works in Washington. In all cases the Committee decided, after careful examination, that the evidence did not justify departure from their main conclusions.

Inherited damage. There have been a number of parallel, and equally important, investigations of the possibility that ionising radiation can cause inherited damage in subsequent generations. The biochemical changes that can follow the absorption of radiation and occasionally lead to cancer may similarly lead to transmissible changes in the reproductive cells of males and females. Such changes may be point mutations affecting single genes, chromosome anomalies or chromosome re-arrangements.

In contrast to the case of cancers, estimates of inherited damage in man due to ionising radiation have to be inferred solely from animal and other experiments. No cases have been proved in any human population. The most extensive studies[7] are those carried out on the children of the survivors of the Hiroshima and Nagasaki bombs – some 30000 children born to irradiated parents were compared with 40000 children born to unirradiated parents without any significant differences in inherited damage emerging. Damage to the foetus has been correlated with irradiation of the mother during pregnancy, but such effects cannot be transmitted to another generation.

Particular studies of genetic traits have been made of populations living in areas of abnormally high background radiation, such as the Kerala area of south west India. This area consists of a thin coastal strip where the beach sand contains monazite, a thorium ore, and the natural background radiation varies from four to ten times that of other areas of India. Evidence of a number of genetic variations has been sought,[12] but the only positive variation found concerned the incidence of Downs syndrome (mongolism), which is due to a numerical chromosome anomaly. The number of cases was higher than in a control population. However, the prevalence rate of Downs syndrome in other areas of India is not known and in Kerala it was not higher than has been found elsewhere. Further, the incidence of the syndrome is known to vary between different areas and at different times for reasons which cannot involve radiation.

A very detailed study [13] has also been made of a population of some 70000 people living in some parts of Guangdong Province, China, where the background radiation level is about three times that in neighbouring areas; a high proportion of the population has lived there for many generations. The populations were compared with respect to overall mortality rates, chromosomal aberrations, and frequency of malignancies. The growth and development of children were compared, as were the occurrence of hereditary diseases and of congenital deformities. No statistically significant difference between the populations living in the high- and low-background areas was found. The frequency of hereditary diseases and congenital deformities as a whole was actually somewhat lower in the high-background areas. The incidence of Downs syndrome was, however, higher (1.7 per thousand) in the high-background area, but it was zero in the control area and the size of population investigated is not enough to establish any statistically significant connection of this condition with radiation.

Estimates of the incidence of radiation-induced hereditary effects in humans therefore rely on evidence from animal studies. Early experiments by Müller demonstrated inherited changes in fruitflies (Drosophila), but the main body of information about the frequency of occurrence in the offspring of irradiated animals has been gained from large-scale experiments with male and female mice. Such experiments have shown that the natural rates of at least five separate types of genetic abnormality are doubled by roughly similar radiation doses and it is currently accepted that the so-called doubling dose is around 100 rem. While extrapolation from animals to humans is associated with major uncertainties, knowledge of the fundamental mechanism of radiation injury at the genetic level is perhaps more complete than knowledge of the mechanism of radiation-induced carcinogenesis, and there is general agreement on the reasonableness of the estimates and on the applicability of the linear no-threshold hypothesis. The consensus view is that the total number of hereditary defects, spread over all subsequent generations, resulting from a given dose of radiation to a population, is of the same order as, and probably somewhat below, the total number of fatal cancers that would occur in the exposed population itself as a result of that radiation.[9] As with cancers, one is looking for a small variation on a fairly common occurrence. Some 30 000 children born in the UK each year suffer from some form of genetic or congenital defect, about 50 per 1000 live births, and it is unlikely that any increase that may result from natural or man-made sources of radiation will ever be observable with any degree of certainty.

The estimates that have been quoted for the delayed effects in exposed individuals and for the hereditary effects are certain to be refined as advances continue to be made in radiobiology, medical physics and biology and as our understanding of the basic mechanisms improves. It is already possible to trace differences in response of different tissues, and it is known that effects in humans vary according to age and sex. Improved knowledge may make it possible to identify groups of individuals who might be particularly at risk. However, the understanding of the biological effects of ionising radiation is good enough now to enable a conservative set of regulations to be drawn up to define the responsibilities of the industry. A recent major US review of research on this subject concluded that 'scientific information is sufficient to permit federal authorities to . . . delineate comprehensive regulatory policies without serious limitations . . . current research on the biological effects of ionising radiation is no longer of vital interest to the several regulatory agencies.'[14]

Effects of chemicals. Before turning to the development of these regulations, however, it is worth noting that the basic problem of low-dose effects is by no means confined to the effects of ionising radiation and to the concerns of the nuclear power industry. A large number of carcinogenic and mutagenic chemical substances have been recognised, and the effects are necessarily established by experimentation at high exposures, usually on animals or simpler organisms. All the problems that we have already discussed in the context of ionising radiation occur again in attempting to extrapolate such results to the delayed effects of low doses on human beings. It is important that the possible effects of low doses of potential carcinogens should be established so that levels can be set for the protection of workers in an industry, and for the general public. The questions again arise: Is there a threshold? What is the shape of the dose–response curve at low doses? The use of a linear dose–response model without threshold for the carcinogenic hazards of polycyclic aromatic hydro-carbons in the urban atmosphere was actually recommended by a Committee of the US National Academy of Sciences in 1972, and a similar relation may be true in the case of asbestos. But to put all this in perspective: Philip Handler, the President of the National Academy of Sciences, has argued in a recent address[15] that the possible effects of all known man-made chemicals taken together could contribute only a miniscule fraction of the total of all carcinogenesis in our population. If carcinogenesis is largely due to environmental factors, he concludes that the environmental culprits remain to be identified and that they will be naturally occurring and widespread.

The uncertainties that remain in the very low figures for risk associated with low doses of a variety of potentially toxic agents are there because the delayed effects are so rare that their measurement is difficult. The uncertainties that remain in the case of radiation are probably less than in most others, because of the very large international effort that has been deployed on this subject. If the effects of low doses of radiation were much larger than the estimates summarised here, the results would have become obvious as significant differences between different populations.

2
The regulation of radiation exposure

Having summarised very briefly the state of scientific knowledge of the effects of ionising radiation, we turn now to the parallel evolution of the principles of radiological protection, which are an attempt to build an acceptable code of behaviour and practice on the basis of these advances in scientific knowledge.

The earliest record that appears to exist of organised steps to avoid radiation damage to people comes from the Deutsche Roentgen Gesellschaft in 1913, who issued an information leaflet calling attention to the hazards of X-rays, making specific recommendations with regard to thicknesses of lead screening to be used and pointing out that workers had the right to refuse radiographic work if the protection arrangements were inadequate. In this country, a motion at the annual meeting of the Roentgen Society on 1 June 1915 read: 'That in view of the recent large increase in the number of X-ray installations, this Society considers it a matter of the greatest importance that the personal safety of the operators conducting the X-ray examinations should be secured by the universal adoption of stringent rules.'[1] The motion was carried unanimously and recommendations were issued in November of that year. Several more countries introduced recommendations in the succeeding years, a process which culminated in the setting up in 1928 of an International Committee on X-ray and Radium Protection, subsequently to be renamed the International Commission on Radiological Protection (ICRP).

One of the most important concepts that was introduced during the 1920s was that of tolerance dose, measurable by means of instruments. The tolerance dose was originally set in terms of a fraction of the erythema dose, the dose required to produce visible reddening of the skin. Mutscheller in Germany and Sievert in Sweden, both in 1925 and apparently independently, suggested that an acceptable exposure for radiation workers was $1/10$ of an erythema dose per year but as physical measurement techniques improved, the tolerance doses were expressed in terms of the ionisation produced in air. At the 1934 meeting of the ICRP in Zurich, a level that could be tolerated by a person in normal health operating under satisfactory working conditions was suggested but it is interesting to note that even in those early days the recommendations included such words as ' . . . should on no account expose himself unnecessarily . . . ' and ' . . . should place himself as remote as practicable from the X-ray tube . . . .' Radium production measures were being developed in parallel.

Information about the genetic effects of ionising radiation was beginning to emerge in the late 1920s and early 1930s but there was not yet enough information to enable this to be taken into consideration in making protection recommendations. However, during 1940 and 1941 there was much discussion at the Advisory Committee on X-ray and Radium Protection in the US on the reduction of the tolerance dose by a factor of 10 to take into account the possibility of genetic injuries. It was clear at that time that the 'tolerance dose' concept might not be applicable to genetic injury since a safe threshold could not be proved and the term 'permissible dose' was introduced so that it should not be implied that no damage whatever was occurring. It was realised, however, that doses could not be reduced by large amounts in practice, while using the equipment currently available, without seriously interfering with the clinical work being done. These matters were not resolved before the Manhattan Project, the wartime atomic bomb project, was launched.

Renewal of international collaboration after the war resulted in the re-formation of the International Commission, the first meeting being held in July 1950 in London. At that meeting recommendations were made on the basis of the large body of information which had by then accumulated, largely as a result of the wartime weapons programme. It was recognised that both carcinogenic and hereditary effects were important and the concepts of maximum permissible concentrations of radioactive materials in air and drinking water were introduced. Maximum permissible exposures were set at a level which involved a risk small compared with other hazards of life. Nevertheless, because of the limitations of the available knowledge it was recommended that 'every effort be made to reduce exposure to all types of ionising radiations to the lowest possible level'.

This caution was reflected in an entirely revised set of recommendations issued in 1958. The Commission based these recommendations on explicit assumptions concerning effects at low doses:

> Any departure from the environmental conditions in which man has evolved may entail a risk of deleterious effects. It is therefore *assumed* that long continued exposure to ionising radiation additional to that due to natural radiation involves some risk . . . the assumption has been made that, down to the lowest levels of dose, the risk increases with the dose accumulated by the individual.

For the purposes of radiological protection, this assumption is expressed as the linear, no threshold extrapolation, as has been discussed in Chapter 1.

In its latest recommendations published in 1977[2] the

ICRP advanced a comprehensive philosophy against which its recommendations might be developed and applied. The ICRP identified three primary objectives:

(i)   JUSTIFICATION: no practice shall be adopted unless its introduction produces a net positive benefit;

(ii) OPTIMISATION: all exposure shall be kept as low as reasonably achievable, economic and social factors being taken into account;

(iii) LIMITATION: the dose equivalent to individuals shall not exceed the limits recommended for the appropriate circumstances by the Commission. The limits† have been set at 5 rem in a year for those occupationally exposed and 0.5 rem in a year for members of the public.

In reaching these conclusions and particularly in deciding on the dose limits, the Commission was guided not only by their assessment of the biological effects of ionising radiation but also by an estimation of what would entail a reasonable or acceptable degree of risk. It is therefore convenient to enumerate here some statistics of the actual risks we run in daily life and in a variety of occupations.

Fatal accident statistics

An 'acceptable degree of risk' is a concept which is very difficult to define; different people will make different judgements of what is most to be feared, and will think of 'risks' in different ways. To equate 'risk' with 'risk of death' is to give a narrow definition, but at least mortality statistics are relatively easily compared and the ICRP based most of their arguments on them. Figure 2.1 shows the mortality statistics for the UK as a function of age, expressed as numbers of deaths per year per million in a given age group.[3] The

---

†These limits are set in terms of whole body doses; ICRP also makes more detailed recommendations on maximum doses for specific organs.

Fig. 2.1. Accident and mortality statistics. Note the scale of ten used on the vertical axis – for example, the total number of deaths per year per million people aged 65–75 is 35 000, compared with a corresponding accident figure of 600.

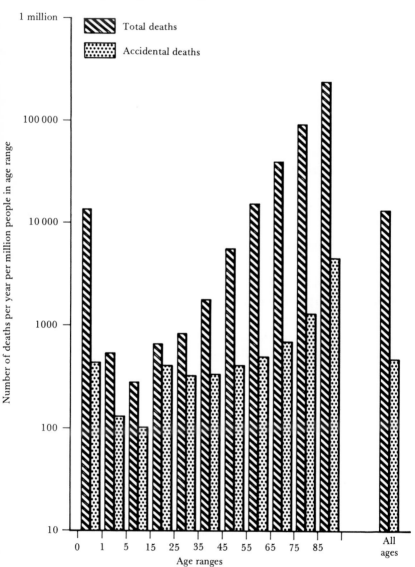

minimum incidence of death from all causes occurs in the 5–14 age group and amounts to 200–300 deaths per year per million children in that age group. The probability of death due to accident also varies with age; the lowest incidence again occurs in the 5–14 years age group, about 100 per million per year. The most common causes are traffic accidents and falls.

It is well recognised that certain groups of people accept, consciously or unconsciously, additional risks of accidental death by engaging in certain activities. The use of road transport should not be classed as an avoidable hazard today but perhaps the choice of riding a motorcycle would fall into this category: an activity 20 times as dangerous as driving a car. Certainly, engaging in sporting activities involves a voluntary acceptance of additional risk of death which may be as high as 1500 chances per year per million people participating in horse, car or cycle racing. If an activity introduces the risk of the development of illness some time in the future, then it is likely that larger risks will be acceptable. The most commonly accepted risk of this type is smoking, where the risk of smoking 20 cigarettes a day may be expressed as accepting an additional 3000 chances in a million of death per year, some time in the future as a delayed consequence.[4] To permit easy comparison of relative risks, all these statistics are expressed in terms of the number of fatalities caused per million of the *exposed* population – that is, per million horse riders or per million smokers, regardless of the actual size of population studied.

The risk associated with various forms of employment is a further indication of the levels of hazard which are associated with present practices. Table 2.1 shows some figures from the Health and Safety Executive Report for 1977. These statistics relate to deaths caused directly by accidents and can only be obtained for industries where the number of accidents is high and the population at risk is large. The statistics will change from time to time as hazards are recognised and remedial measures are adopted. The improvement in the safety of coal mines is a good example – 1279 miners from a mining population of 918 600 (one in 718) were killed in the UK in 1907 compared with 34 from a mining population of 218 500 (one in 6426) in 1981. It is more difficult to obtain statistics on deaths caused by diseases contracted while at work. There was early interest in the damage caused to miners' lungs by dust, and Case and Doll pioneered studies of rubber and chemical workers in the 1950s. Some figures quoted by Pochin in 1975[5] are shown in Table 2.2, though they do not refer necessarily to current levels of occupational exposure.

It may be seen from Table 2.1 that the UK fatal occupational accident rate ranges from about 20 to 100 per million per year, in a whole range of occupations normally considered safe, such as the manufacture of textiles, paper, furniture, bricks, chemicals, metals and shipbuilding. Occu-

Table 2.1. *Industrial risks – deaths from accidents*

UK industry	Deaths per million employees at risk per year
Textiles	20
Manufacturing industries	34
Chemicals	50
All industries (average)	67
Shipbuilding	100
Coal mining (underground workers)	180
Quarrying	240
Fertilisers	410
Coke ovens	420
Deep-sea fishing	2800

Source: Health and Safety Executive

pations normally rated as more hazardous, such as quarrying and coal mining, have accident rates of up to 300 per million per year. The Commission bases its recommendations for occupational exposure on comparing the risk that might arise on average in radiation work with that for other occupations recognised as having high standards of safety. In such industries the average annual mortality due to occupational hazards was thought not to exceed one in 10 000, or 100 per million per year, an assumption which is supported by these statistics.

Limits of radiation exposure

In recommending a limiting maximum whole-body dose of 5 rem a year for workers with radiation, the ICRP assumed that the need to keep within this limit and observance of the other procedures that were specified would, in practice, reduce the average exposure of occupational personnel to annual doses of about 0.5 rem, which is between

Table 2.2. *Industrial risks – deaths from disease*

Industry	Chance of death per year per million employees at risk
β-naphthylamine (bladder cancer)	24000
Rubber mill workers (bladder cancer)	6500
Underground mining (pneumoconiosis and silicosis)	140–5800*
Viscose spinners (heart disease)	3000
Coal carbonisers	2800
Asbestos (lung cancer)	2300–4100
Uranium mining (lung cancer)	1500
Wood machinists (nasal cancer)	700
Shoe industry (nasal cancer)	130
Range for accidental deaths	34–2800

*Large differences in rates for different countries and different types of work.

two and three times the background dose. The average annual dose received by workers at nuclear power stations in the UK is actually about 0.3 rem, and the average annual dose received by fuel reprocessing workers is about 1.2 rem.[6] In the US, the average for all employees in the nuclear industry is 0.36 rem.[7] The assumption of an average annual dose of 0.5 rem, by the ICRP, therefore seems justified. Based on the figures for somatic effects used by the ICRP, this will give a figure of excess cancer mortality of about 60 per million per year; and roughly the same number of hereditary defects may occur. (The actual number of cases would be much smaller because the population exposed is small.) If the same individual accumulated this dose every year over a working life of 40 years, his chance of eventually dying of cancer would increase from about 0.200 to 0.202, an increase of 1%. By comparison with the figures for other industries, it may be seen that the radiation hazards suffered by people in the nuclear industry are statistically comparable with the accidental risks, and probably well below the risks of death from occupational diseases in industries not normally considered to be hazardous.

To turn now to a matter of perhaps greater interest: the protection of the general public. The Commission's recommendations are based on the selection of appropriate critical groups who are likely to be the most exposed as a result of any particular operation. Such a group should be representative of those individuals expected to receive the highest dose. The Commission again takes a view of an acceptable level of risk for members of the general public, and concludes that the general level of acceptability for fatal risks is a factor of 10 lower than for occupational risks. They therefore set the maximum dose to the most exposed individual as 0.5 rem in any year, and they suppose that the application of this annual dose limit to individual members of the public is likely to result in average doses of 0.05 rem a year, considering the maximising assumptions usually made in selecting critical groups and the fact that exposures at the limit are not likely to be repeated every year for many years. The exposure limit, 0.5 rem per year, is less than three times the average natural background in the UK and is approached by the natural background in some other parts of the world. In fact, the average radiation doses to the public resulting from the nuclear industry are very much lower.

All discharges of radioactivity are regulated in the UK by the Environmental Departments and the Ministry of Agriculture, Fisheries and Food (MAFF). The latter publishes an annual survey of radioactivity in surface and coastal waters[8] and the National Radiological Protection Board (NRPB) publishes calculations of the annual radiation exposure of the UK population from discharges of radioactive effluents of all types.[9] The latest MAFF report published in 1983 shows that all emissions from all nuclear installations were within the authorised levels and that no member of the public has received a dose exceeding the ICRP limit. For the great majority of installations, indeed, the most highly exposed member of the public received less than 1% of this limit, whilst some groups were exposed to doses of between 1 and 3% and one small group to 69% of the ICRP limits.

Results for the irradiation of the general population published by the NRPB show that over three-quarters of the total radiation we all receive comes from the natural background, that is, from cosmic rays, terrestrial γ-rays, internal radiation, and radon decay products. These results are illustrated in Figure 2.2. Radiation from medical uses far outweighs any other source of man-made radiation and, indeed, emissions from the nuclear industry are, in total, the lowest of any source monitored, amounting to $3/10000$ rem per person. On the basis of the usual linear extrapolation model, therefore, the radiation dose due to the operations of the nuclear industry might account for one excess case of cancer per year in the general UK population of 56 million and, in time, in a similar number of hereditary defects. The confidence expressed by the Commission that the application of their recommendations to critical groups would lead to very low exposures of a large population is amply justified.

Fig. 2.2. Average annual radiation exposure of the population of the UK.

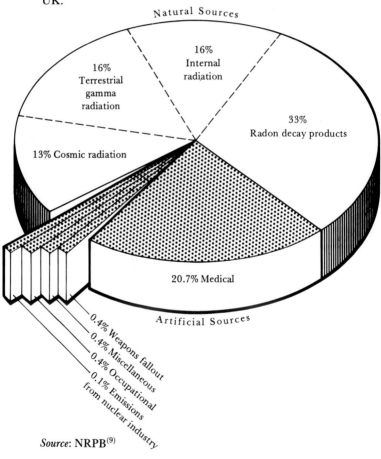

Natural Sources

16%
Internal
radiation

16%
Terrestrial
gamma
radiation

33%
Radon decay products

13% Cosmic radiation

20.7% Medical

Artificial Sources

0.4% Weapons fallout
0.4% Miscellaneous
0.4% Occupational
0.1% Emissions
from nuclear industry

Source: NRPB[9]

The estimation of benefits

A system of maximum-dose limitation is only part of the general philosophy of radiological protection which has evolved in the ICRP and in parallel national bodies. The first principle recommended by the Commission is that: 'No practice shall be adopted unless its introduction produces a net positive benefit.' It is relatively simple to apply such a principle when one is comparing like with like and where the benefits will accrue to the same group that bears the possible detriment. Perhaps the most clear-cut example is the estimation of the net benefit to be gained by the radiological screening of healthy people, because of the risk that exposure to diagnostic doses of radiation might cause more deaths from cancer in the long term than they prevent. Studies of this sort have been made of the relative advantages of screening for breast cancer and cancer of the stomach and also of thyroid scanning; it is evident that, in all such cases, the hazards decrease with age. Any such studies must, of course, take account of the uncertainties of extrapolation to low doses, which we have already discussed, and of the need to consider the risk of hereditary defects. Indeed, the NRPB has recently argued that, because of this, medical techniques should be reviewed so that doses received by patients, which vary widely for similar treatments, should be reduced to the minimum applicable to the particular use.[10] Inspection of Figure 2.2 shows that action to reduce medical irradiation doses would be by far the most efficacious in reducing the general radiation dose to the population – if that is considered to be a worthwhile aim.

The principle of achieving positive net benefit is much more difficult to apply when the benefits are general and the risks are concentrated in critical groups. Indeed, this same question might be raised whenever there is any suggestion of introducing a new technology, or even a new product, in the form of: 'Will the general benefits of this new introduction justify the occupational hazard of those engaged in producing it?' So one needs an agreed unit in which both benefit and harm might be expressed. Perhaps one way in which such a general question can be framed relative to the advantages or disadvantages of the nuclear generating industry as a whole is the following: Having regard to the estimated risks on the best-available knowledge, is the generation of electricity in nuclear stations safer or less safe than alternative methods of achieving the same end? That question has been addressed by many authors.[11, 12] It is straightforward enough to assemble statistics of the number of fatal accidents associated with the winning of an equivalent amount of fossil fuel and its uses in the generation of electricity and to compare that with the record and possible radiation damage arising from the nuclear industry, taking into account all stages of the nuclear fuel cycle, including the mining of uranium. In this last occupation, the occupational risks are similar to those of other dangerous occupations, but the overall weighting is small because of the relatively small quantities of uranium that have to be mined to generate a given amount of power.

A difficulty arises when we try to make a similar comparison for the impact on the public. Despite many uncertainties, a reasonable scientific basis has evolved, as we have seen, to enable such an estimate to be made in the case of nuclear power and the nuclear fuel cycle. It seems that as good a basis does not yet exist for the estimation of the impact on the health of the public of burning fossil fuels. We again face the problem of the estimation of the effects of small doses of substances known to be harmful in large concentrations. The combustion products of fossil fuels include sulphur dioxide, much particulate matter, appreciable quantities of radioactive substances, and known carcinogenic and mutagenic compounds such as benzopyrene. Estimates of the effects have varied widely, from about two to 40 deaths per GW-year of electricity – the annual output of a large power station. Taking into account the possibility of accidental deaths as well, Sir Edward Pochin[11] estimated in 1980 that the total number of deaths due to one GW-year of electricity output might be approximately 15 if the fuel source is coal; ten if the fuel source is oil; and two if the fuel source is nuclear. Cohen & Pritchard[12] of the Health and Safety Executive recently published a critical survey of the many studies that have been carried out on the comparative risks of electricity production systems. They conclude that suitably sited, constructed and maintained nuclear systems would involve not more, and probably less, risk than oil or coal burning systems, taking into account in each case of the whole fuel cycle; they add: 'We doubt if further comparative studies would greatly refine this conclusion.' The overall result of these studies is that considerations of safety either to workers or the public need not limit the scale of the use of nuclear power, and the environmental advantages of the use of nuclear power can be seen as a general benefit.

The ALARA principle

To turn now to the remaining principle of radiological protection, enunciated by the International Commission: 'All exposures shall be kept As Low As Reasonably Achievable, economic and social factors being taken into account', which has become known as the ALARA principle. The Commission acknowledges that radiation risks are a very minor fraction of the total number of environmental hazards to which members of the public are exposed. The Commission is here advocating a system of dose optimisation, ideally determined by cost–benefit analysis, the purpose of which would be to ensure that the total detriment should be appropriately small in relation to the benefit resulting from the introduction of any proposed activity. In order to make any such comparisons it is necessary to express costs and benefits in the same units. Costs can be expressed only in money terms. From the point of view of radiological protection, benefits are seen in the reduction of collective dose or man-rems of exposure, that is, the sum of the doses to all exposed individuals; for example, 10 000 people each receiving $\frac{1}{10}$ rem corresponds to 1000 man-rems. To compare the two we must have some rational basis for the estimation of the value of the man-rem and, furthermore, this must be expressed in terms which will be a help to practical decision making. Ascribing money values to risks or detriment is always difficult since one cannot express all possible harm to human beings in quantitative terms. The result of any dose optimisation procedure can therefore only be one input into a final decision on what degree of protection is appropriate, and all such procedures must be used with caution.

Nevertheless, carrying out cost–benefit analyses yields at least one means of making comparisons between different procedures and different situations. It is already obvious that one component in any such process of optimisation will be the value to be ascribed to very small increments of dose that may nevertheless affect large numbers of people. Webb & McLean[13] of the NRPB introduced, in 1977, the concept that there exists a level of annual dose which is insignificant to the individual on the basis of an analysis of risk from other causes. They proposed that an annual dose of $\frac{1}{100}$ rem ($\frac{1}{20}$ of background) could be considered as insignificant. Such a dose would give an annual risk of death of about one in a million, not much larger than the risk of being killed by lightning and equivalent to the risk run voluntarily by someone smoking $1\frac{1}{2}$ cigarettes in a year, again on the assumption that there is no risk threshold for that particular activity!

This dose is much less than the variation in natural background, which we do normally neglect as being of no practical significance to health; no-one cancels a trip or a move to Aberdeen or to Cornwall because of the relatively high background radiation in those localities, which results from the high uranium content of the local granite.

The most recent development is the publication of a report by the NRPB on Cost Benefit Analysis in Optimising the Radiological Protection of the Public: a Provisional Framework[14] which had been preceded by the circulation of a consultative document. In this report, it is argued that while, on the basis of the linear hypothesis, risk is directly proportional to dose, it does not follow that the costs which might be associated with averting risk should also be directly proportional to dose. The Board suggests that individuals require increasing compensation for exposure to increasing risk, and it seems reasonable to adopt a dose optimisation procedure which will concentrate protective resources on the individuals likely to be at the highest risk; this approach is consistent with the ICRP's concentration on the protection of a critical group of the most highly exposed individuals, since average risks are so low. The tentative values suggested for a man-rem in different ranges of dose are shown in Table 2.3; given that 10 000 man-rem may result in the loss of one

Table 2.3. *Proposed values of man-rem at different dose levels*

Dose level expressed as percentage of ICRP limit (0.5 rem per year)	£ per man-rem
<1	20
1–10	100
10–100	500

Source: NRPB

life, these values range from £200000 per life saved at the lowest doses to £5 million per life saved at doses approaching the maximum allowed – as we shall see this is a high figure compared with the money spent in avoiding other risks. At the lower end of the dose range, the significance of collective doses to populations when the individuals comprising the population are exposed to less than 1% of the ICRP limit is open to doubt; no individual detriment has been proved at such low doses.

Evolution of science and regulatory processes

We shall return in the next chapter to the concept of acceptable risk and the costs that an industry may be expected to pay for safety. It is appropriate to consider here the conditions under which the science and regulatory processes we have been following can advance. The theory of radiological protection has evolved from the first simple idea of setting exposure levels at which obvious symptoms could be avoided, to encompass a wide range of very complex scientific questions which, in the low-dose range of practical interest, involve matters that are beyond direct experimental proof and depend upon a very careful and very detailed interplay of theoretical approach and practical experimentation on model systems far removed from man.

Two stages can be discerned in the evolution of such a subject. The first is the reporting of new experiments on radiation effects, the assembly of fresh epidemiological data and so on. Papers on these subjects should be published in the scientific literature, and be open to the processes of peer review, tests for internal consistency and logic, comparison with other results and repetition under the same conditions wherever possible, which are the normal and well-understood methods by which scientific knowledge is advanced. The second is the digestion and evaluation of a large amount of diverse results in a consensus view of the best advice that can be given on subjects which are complex and which cannot be settled by direct experimentation. Such questions can only be resolved by the painstaking and detailed exercise of scientific judgement and one therefore has to create an organisation of scientists drawn from a range of disciplines in which results can be assessed and a reasonable consensus reached, calling on independent experts who are competent to discuss particular branches of the subject.

It is fortunate that scientific organisations of great weight and authority have been formed to review the subject of radiological protection. The International Commission, for example, consists of members chosen on the basis of their recognised expertise rather than on a national basis. The Commission is independent of national institutions and is supported by grants from organisations such as the Ford Foundation, the World Health Organisation and the United Nations. The United Nations Scientific Committee on the Effects of Atomic Radiation was established by the General Assembly of the United Nations in its tenth session from 15 member states. Again, the Committee on the Biological Effects of Ionising Radiation was set up in the US by the National Academy of Sciences and consists mainly of scientists attached to US Universities and Medical Schools. The status of the Medical Research Council in the UK, with its various expert committees, is well known. The degree of consensus reached by these various organisations is impressive. That institutions of this degree of independence are active in the field is not to say that controversy will not continue, nor that unexpected results and new insights will not occur. It would also be naive to think that scientists engaged in exercises of this type are unaware of the political pressures that surround every aspect of work that has, as its ultimate aim, the protection of the public. But the sort of scrutiny that published results receive from the scientific community should be perfectly adequate to guarantee that a false consensus will not last long. One must feel sympathy for a scientist who believes that some new results show the possibility of a greater public hazard than hitherto realised and who feels that his/her obligation as a citizen demands that wide publicity be given to them. But the discipline of full publication in the scientific literature, of peer review, and of repetition by other observers (where possible), should not be bypassed. To quote again Dr Handler,[15] the President of the National Academy of Sciences, in 1979:

> What seems lost on some who would participate in the debate on the place of technology in our society, particularly those concerned with possible environmental carcinogenesis by radiation or chemicals, is that the necessity for scientific rigor is even greater when scientific evidence is being offered as the basis for the formulation of public policy than when it is simply expected to find its way in the market place of accepted scientific understanding. Science itself can benefit by early publication of properly documented preliminary findings. But surely public policy should not rest on observations so preliminary that they could not find acceptance for publication in an edited scientific journal. And yet that has happened repeatedly.

> Political decision-makers have no choice but to rely on the validity of what seems to them to be the findings of rather recondite science thereby placing a heavy onus on scientists who bring such matters to attention. Announcement in the press of each experiment, in turn, generates public alarm that can neither be justified nor assuaged.

If the scientific ideas we have been following have to evolve in a scientific context, their translation into industrial regulations demands other institutions. The ICRP has indeed gone beyond the scientific question of the biological–medical effects of radiation in promulgating recommendations and general rules based on the philosophy of comparison with other industries, that is, they have made a value judgement. Whether these rules and comparisons are acceptable and how they are to be applied is no longer a purely technical matter. In this country, the NRPB exists to weigh the evidence and to issue operational advice. At that level, value judgements have to be made and explicitly seen for what they are, though the arguments can still be objective. Finally, there need to be methods of proper debate with bodies accountable to the public, and the responsibility for final decisions on acceptable levels of risk is a matter for political decision. What is important is to achieve the discussion and evaluation of different classes of ideas within appropriate organisations, otherwise much unnecessary and damaging confusion arises.

3

The risk of accidents and the cost of safety

In the first two chapters, we have discussed the possible radiation hazard arising from the routine operation of the nuclear industry, and followed the evolution of the scientific basis for the system of control which is currently applied, along with the parallel development of the regulatory framework. But control of normal operation is only one of the responsibilities of industry. There is also the requirement to reduce the probability of accidents affecting the public or workers to levels which are deemed to be acceptable. This chapter will therefore be concerned mainly with the risk of accidents. We shall take as a prime example the types of accident to which nuclear reactors may be subject, and consider the development of methods of risk assessment which can be applied to them and, indeed, to complex plant of any description.

As a starting point it is necessary to define the term 'risk'. In engineering or scientific terms, risk is expressed as a combination of the probability of an occurrence and the magnitude of its consequence, though the popular interpretation of the word 'risk' may be different and is probably very subjective. We shall first discuss the evolution of the relatively new branch of engineering known as *risk assessment* – the determination of the product of probability and consequence – then describe the evolution of criteria that may be used in judging the acceptability of the assessed risks, and finally discuss the complementary subject of the perception of risk, and how this perception influences the resources devoted to its avoidance. A much more detailed treatment of the subject has recently been published by the Royal Society.[1]

Risk assessment

The simplest approach to risk assessment – the historical approach – is based upon past experience. Such an approach is only possible, however, when the experience is extensive – for example, in assessing the effect of different methods of road construction on traffic accidents, or comparing the safety of different types of bridges. In 1957, when talking about nuclear power, Christopher Hinton said that: 'All other engineering technologies have advanced not on the basis of their successes but on the basis of their failures.'[2] As the scale of technology increases, the potential damage which could result from a major accident in many industries – not only in the nuclear industry – is such that progress through learning the cause of failures is becoming less and less acceptable.

It was realised from the outset that accidents involving loss of life of members of the public were possible at nuclear plants. However, the remarkable safety record of the nuclear industry has meant that the historical approach for risk assessment cannot, in general, be used. Therefore, the practical approach to reactor design assumed the same sort of philosophy that is commonly applied in engineering to the design of plants and large structures; their safe operation is based on the calculation of the maximum expected event (for example, maximum load) and the structure is designed to resist that event with a due safety margin; it is then deemed to be safe for minor events. The approach is described as designing to survive the 'design basis event'. However, one can clearly imagine events of greater severity than the design basis event. Such events are not 'impossible', but they are extremely unlikely, indeed, we shall see that one can calculate the probability of such events occurring and that this probability becomes almost vanishingly low as the postulated event becomes more and more severe.

The point may be illustrated by a non-nuclear example. It is possible that two jumbo jets could collide in mid-air; it has never happened, though such a collision has occurred on the ground. Six hundred people would probably die as a result of a mid-air collision; the worst recorded aircraft accident actually caused 581 deaths. However, this accident might occur during the approach to an airport near a big city, say, London. One aircraft might fall on a petrochemical works (such as Canvey Island) initiating damage which could lead to a further accident, perhaps killing thousands of people. Or an aircraft might fall on a crowded football stadium, and the total result might be a major disaster affecting tens of thousands of people. One seldom takes any account of such unlikely events, even though they are not physically impossible, simply because the probability of such an event is low. 'What if this and then this and then this should happen . . . ' is the language of phobia, not of common sense.

Up until the mid-1960s, it was generally felt that insufficient data were available to carry out predictive calculations on plant as complex as nuclear reactors. Professor F.R. Farmer introduced a new and rational approach in his classic paper 'Siting Criteria for Nuclear Reactors' published in 1967.[3] Following the pioneering work of Farmer and others, it became clear that the many possible events that can occur in a complex plant can be broken down in a systematic way into the behaviour of simpler pieces of equipment and the results of individual operator actions. Nuclear plants, like all conventional plants, consist in the main of standard pieces of equipment – pumps, valves, motors – and it is possible to obtain data on the reliability of these pieces of equipment, using historical data from a wide variety of sources. Large data banks on the reliability of components, such as have been built up in the Safety and Reliability Directorate of the UKAEA or in the EEC Laboratory at Ispra, contain records of thousands of individual items. The performance and reliability of complex systems of these components may then be predicted.

The systematic technique available for analysing the behaviour of complex plant is known as 'event tree analysis'. As an example of an event tree, let us consider a possible sequence of events, illustrated in Figure 3.1. An initial event

Fig. 3.1. Simplified example of an event tree sequence following a feedwater pump failure.

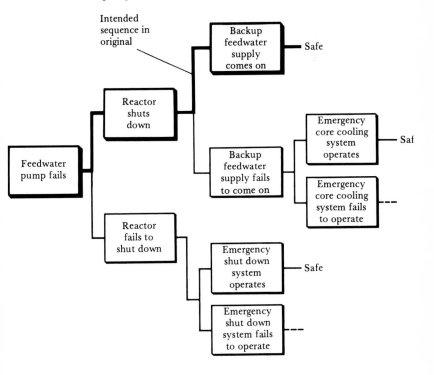

could be the failure of the pump supplying feedwater to the steam generators (boilers) on a nuclear reactor. Without a supply of water to the boilers, heat can no longer be removed from the reactor and the reactor will start to heat up. Such an eventuality would have been foreseen, and safety devices built in to avoid serious consequences and eventually to return the plant to a safe condition. At each stage, these may work satisfactorily, or they may not. By assigning a frequency to each initiating event, and then probabilities to each subsequent stage, it is possible to arrive at a probability for each of the possible consequences. In a complete event tree analysis, all events that can be foreseen, at all levels of probability, are considered. By carrying out an assessment of every potential initiating event and postulated failure, and of their consequences, the reactor designer can build up a probabilistic representation of the risk of the entire plant.

Carrying out a probabilistic assessment of risks enables the reactor designer or safety assessor to define a series of accident sequences of decreasing probability. There is no need to decide which sequences are 'possible' or 'credible' and which are not, but it is eventually necessary to decide at what frequency of occurrence a given sort of accident might be tolerated. These 'acceptance criteria' are discussed below. In fact, the value of carrying out risk calculations in a thorough and systematic way is as much qualitative as quantitative. The analysis provides a framework in which a safety assessor can ask the question: 'What if this or that happened . . . ?', and a rational answer of probability and consequence of this new possibility can be produced. Again, carrying out the calculations may highlight certain sequences of events where additional safety equipment is required, or show that it would be of little value to improve the reliability of equipment to protect against accidents which were already adequately protected against.

Characteristics of reactor accidents

Before discussing the results of this type of analysis, it is necessary to consider the types of reactor accident that could occur and those that could not. A reactor cannot explode like an atom bomb; such an enormous energy release can only be achieved by forcing together a greater than critical mass of pure or nearly pure fissile material, and the physical arrangement and low fissile content of reactor fuel prevents any such event. However, it is possible for the power in a reactor to exceed the design level and for the fuel to over-heat. Nuclear reactors therefore include safety features designed to prevent loss of control of the nuclear reaction and to maintain cooling on the fuel elements. Should the ordinary control rods fail to act, the control of the nuclear reaction is maintained by incorporating emergency means of inserting neutron absorbing material into the reactor core automatically. Considerable diversity and redundancy is built into these protective systems, and also into the emergency systems for cooling the reactor core if, for any reason, there is an interruption or loss of the main coolant flow.

A loss of cooling is particularly significant because, although the principal nuclear reaction can be rapidly terminated by inserting material to absorb neutrons and hence stop the neutron chain reaction, it is not possible to stop the generation of heat associated with the decay of the fission products. Immediately after shutting the reactor down, this decay heat is about 7% of the original heat output; it then rapidly decreases, reaching about 1% of the original level after an hour and 0.1% after six months. Therefore, unless adequate cooling is provided to remove the decay heat, both immediately after shutdown and for a considerable time afterwards, the fuel elements may heat up and become damaged, leading to the release of some of the radioactive fission products in the fuel.

Nuclear reactors are therefore designed both to prevent over-heating, and with multiple barriers to reduce the chance of radioactive materials escaping if both the main and emergency cooling systems should fail. The first barrier is the form of the fuel itself since the fission products are created throughout the fuel and must migrate to a free surface before they can escape. The volatile fission products, gases like krypton and xenon, together with iodine and caesium, are the most significant. These fission products are further contained by the next barrier – the cladding which surrounds the fuel. An important aim of the safety system on all reactors is to maintain the integrity of the fuel cladding by limiting its temperature. If the fuel cladding should fail, the next barrier is the circuit containing the primary coolant. In all modern reactors this circuit is then surrounded by a pressure resisting containment structure, and in some reactor designs the primary containment is supplemented by a secondary containment building which acts to force any released material to pass through filters before being released to the atmosphere.

Because of these multiple barriers to any uncontrolled release of radioactive material, a nuclear reactor can suffer serious damage within the plant without any major hazard being caused to the surrounding population. The design incorporates a number of emergency systems aimed at preserving the integrity of one or other of these barriers, engineered to sufficient standards of reliability and with the necessary degree of duplication and redundancy, so that the probability of the barriers being breached is reduced to low levels. The safety of the plant may be described by a set of calculations of the probability of radioactive releases of varying magnitude. The protection of the public against the consequences of reactor accidents therefore consists of protection against irradiation or the ingestion of radioactive substances;

reactor accidents are not characterised by extensive off-site damage, as in a fire or explosion, though there could be an expensive clean-up operation if material such as radioactive caesium manages to escape and becomes deposited on the surrounding area.

The second part of a complete reactor risk assessment therefore consists of the calculation of the consequences of various releases of radioactivity. The radiation dose to which people might be exposed depends on the size and type of release and the weather conditions that determine how the active material is dispersed or deposited; the collective dose depends on the distribution of population near a particular site. Meteorological data can again be injected into a probabilistic type of analysis, which will lead to a best estimate of the most likely consequences. Any such analysis has to take into account the efficacy of simple remedial measures, particularly in limiting exposures, while a radioactive cloud is passing – by sheltering indoors, or by evacuation of the population downwind from the reactor. In the UK every reactor or nuclear site operator is required to draw up emergency plans which can be put into effect for the protection of the public if an unplanned release occurs. The levels of radiation dose at which public authorities are required to consider protective measures, known as Emergency Reference Levels (ERL), have been specified by the National Radiological Protection Board. They are set on the basis of a comparison between the risks associated with the radiation exposure and the risks associated with the measures to prevent that exposure. For example, only above a dose of 10 rem (to the whole body) does the NRPB recommend that evacuation should be considered. However, the simpler and essentially risk-free measure of sheltering is worth considering when the likely dose is as low as 0.5 rem. The ERL at which evacuation is to be considered is well above the limit set for prolonged exposure of the most exposed group of the general population under normal operating conditions (0.5 rem per annum), for the simple reason that one is considering very rare events. A dose of 10 rem is approximately equal to the life-time dose to an individual from background sources and corresponds, on the basis of the conservative linear hypothesis, to a possibility of one in 1000 of developing cancer during the 40 years following the exposure.

The results of risk assessment of nuclear reactors

The assessment of the probabilistic risk of a nuclear reactor on any given site therefore requires the calculation of the probable frequency of any given accident sequence, the likely release of radioactivity that would result and the likely effects of such a release, again having regard to the fluctuation of meteorological conditions around the reactor site. A single answer cannot be expected: all such calculations are expressed as a range of values. The summation of the frequency – consequence curves for all types of accident gives an overall 'risk curve' for a certain type of plant on a given site. The first complete study of this type to be published was an assessment of accident risks in US commercial power plants (the WASH 1400 Report) made in 1975 by a study group led by Professor N. Rasmussen of the Massachusetts Institute of Technology.[4] The group concentrated on accidents that could involve melting of the fuel, since these were the only accidents that could involve serious consequences for the public owing to a possible large release of radioactivity. Even so, they concluded that melting of the fuel does not necessarily result in an accident damaging to the public; a spectrum of accidents can occur. They took into account unlikely external initiating events – tornadoes, floods, aircraft crashes and earthquakes, as well as malfunctioning of the reactors. The study concentrated on two US reactors, a pressurised water reactor at Surry, Virginia and a boiling water reactor at Peach Bottom, Pennsylvania. The results were extrapolated to refer to a nominal programme of 100 large reactors; the authors considered this to be a pessimistic approach, since no credit was taken for advances in safety technology. The results were expressed as numbers of fatalities occurring for events of a given frequency and the estimates for 100 nuclear power plants were compared with those for man-made and natural disasters. The probability and numbers of fatalities that were calculated for 100 reactors are shown in Figure 3.2. The information plotted for non-nuclear accidents was taken from the *Statistical Abstracts* for the US for the year 1969.

The Rasmussen Report has been examined by many bodies and much debated; a review group was set up by the Nuclear Regulatory Commission under Professor H.W. Lewis, which issued a Report in 1978.[5] Reviews were also carried out by the Electrical Power Research Institute of the US, amongst many other organisations, and in 1979 they published a comparison of the EPRI and Lewis Committee Review of the Reactor Safety Study.[6] The Lewis Committee endorsed the methodology followed by Rasmussen and the use of probabilistic techniques as the most appropriate for the assessment of reactor safety, but thought that the accuracy of many of the probability factors calculated in the WASH 1400 Report were not as good as claimed and that the resulting range of values should have been stated more clearly in the Report's conclusions. The EPRI team went through many of the calculations again and decided that the median figures in WASH-1400 were probably too high, overstating the risk, although the error band – the range of values quoted – was too small.

Another comprehensive study of the safety of pressurised water reactors was carried out for the German Federal

Fig. 3.2. Actual and estimated probabilities of man-made events involving fatalities.

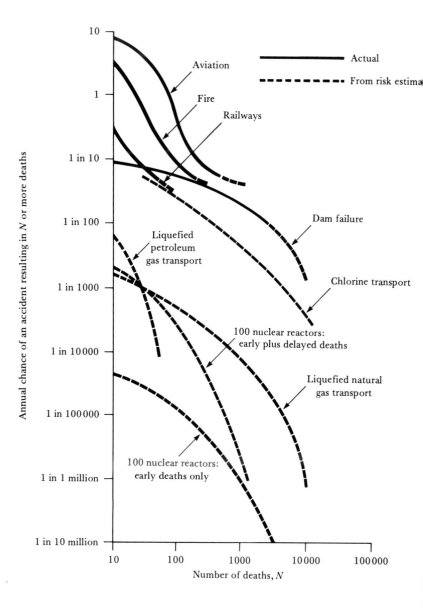

Ministry for Research and Technology, by a team including representatives of many institutes in Germany and led by Professor A. Birkhofer. This study, known as the German Risk Study, reported in August 1979.[7] The study aimed to apply the general methodology and approach of the US studies just mentioned to 25 large PWRs on German sites. Comparison of the German and the US results shows no great differences; certainly no greater than the range of uncertainty which must be associated with both of them. Further studies of the same general type have been published concerning the PWR at Zion, Illinois, and the proposed PWR at Sizewell in the UK.[8, 9] An approximate comparison of the overall results – probabilities of early fatalities – from these four studies is given in Figure 3.3. The curve taken from the German Risk Study was scaled down to one reactor and therefore represents a 'composite' result, and that for the Reactor Safety Study was for a PWR on the Surry site.

The four curves in Figure 3.3 should not be used to make any quantitative comparison of the safety of the plants studied. Any such comparison would have to start with an exhaustive study of the statistical spread of the results in each case and a comparison of the precise assumptions. Furthermore, the cases are not strictly comparable: the population distribution is different in each case; the curve from the German study is a composite one and the results for the Zion and Sizewell reactors exclude external events such as earthquakes and aircraft crashes. The important point is the very low probabilities of any serious accidents, arrived at in four independent studies. Nevertheless, it is probably significant that the more recent studies (Zion and Sizewell B) yield even lower values of these probabilities, since improvements in safety measures continue to be made, and the experience on which probabilities of failure are calculated also improves with time: since worst values have to be

Fig. 3.3. Estimated probabilities of reactor accidents involving multiple early fatalities. The curves show the estimated probabilities (per reactor year) of accidents resulting in more than a given number (X) of early fatalities. For example, for Sizewell B, for each year of operation, point A shows the chance of an accident killing ten or more people and point B shows the chance of an accident killing 100 or more people.

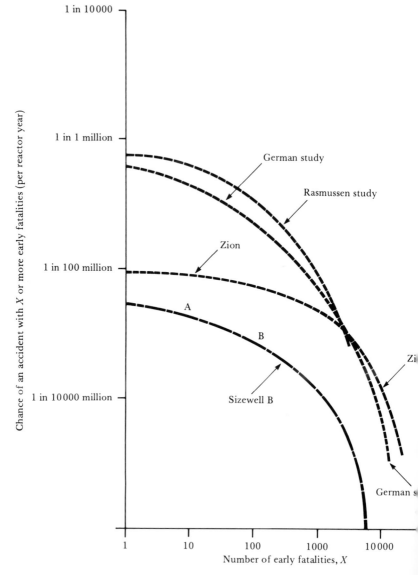

assumed in this type of analysis when data are sparse, the judgements made tend to become less conservative as more data become available.

As well as calculating the numbers of people at risk of early death from high radiation doses, the assessments have been extended to calculate the numbers that might be subject to delayed effects, based on the statistics of the long-term health effects discussed in Chapter 1. Calculations of the probabilities of eventually causing numbers of fatal cancers due to reactor accidents in the four cases already considered are collected in Figure 3.4.† The frequency and number of potential cancers calculated in the German Risk Study were larger than Rasmussen estimated in WASH 1400, partly because of the higher population density and partly because of the different conventions used in the calculation of casualties in very large populations being subjected to small or very small radiation exposures. The German authors state that, except in the most serious and least probable accidents, the majority of all calculated late fatalities resulted from radiation doses of less than 5 rem per individual, which is less than the radiation dose received from the natural background in a life-time. As we have seen, the actual health impact of such low radiation doses may be much less than that predicted on the basis of the linear hypothesis of radiation dose and effect which was used in these calculations.

The results of this type of risk assessment show that accidents to nuclear reactors, causing any casualties at all, will be rare events, and that all but 1% of those that may occur will cause no early deaths and will not be comparable with disasters like earthquakes or explosions in which many people in the same neighbourhood are affected. Any health effects that may follow an escape of radioactivity will be

†Figures 3.3 and 3.4 were compiled from the calculations published in references (5, 6, 7, 8, 9) by Mr John Ward of AERE Harwell.

Fig. 3.4. Estimated probabilities of reactor accidents involving cancer deaths. The curves show the probabilities (per reactor year) of accidents resulting in more than a given number (Y) of cancer deaths.

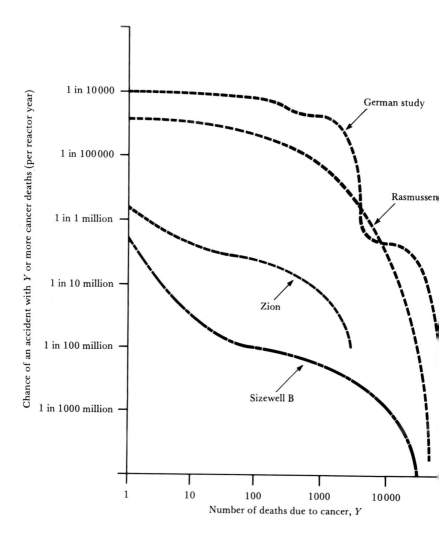

spread through a large population in a random manner, as is the case in many types of accidents involving individuals or for the health hazards of smoking.[10] However, the probabilities of occurrence of health effects due to reactor accidents is very much lower than the incidence of common accidents or of disease due to smoking. To illustrate this point, the German Risk Study[7] included a comparison of the number of reactor accident-related fatalities expected in any year (equal to the number expected to follow an accident multiplied by the probability of the accident occurring in a year) with the number of deaths due to leukaemia and cancer in the same population in a year, and with the number of deaths in that population caused by natural background radiation in a year, calculated on the same conservative assumptions of the health effects of low doses of radiation as those used in the German reactor study. This comparison is given in Figure 3.5, where the probability of individual fatalities is plotted against the distance from a single plant. The estimated numbers of deaths from reactor accidents, when reduced to an average number per year, are very much less than the estimated numbers due to the effects of natural background radiation, and very much less again than the actual number of cancer deaths in that population. We are dealing with events of almost vanishingly low probability.

The Three Mile Island accident

Although many minor accidents to reactors have been recorded, the most severe accident to a commercial nuclear power plant, and the only one involving damage to the reactor core with consequent release of radioactivity, occurred on 28 March 1979, to Unit 2 of a twin 900 MW PWR station at Three Mile Island, near Harrisburg, Pennsylvania (TMI). It is an example of a loss of coolant accident causing severe damage to the reactor core, but actually a negligible

Fig. 3.5. Comparison of estimated effects of major reactor accidents with estimated effects of natural background radiation and actual cancer death rates.

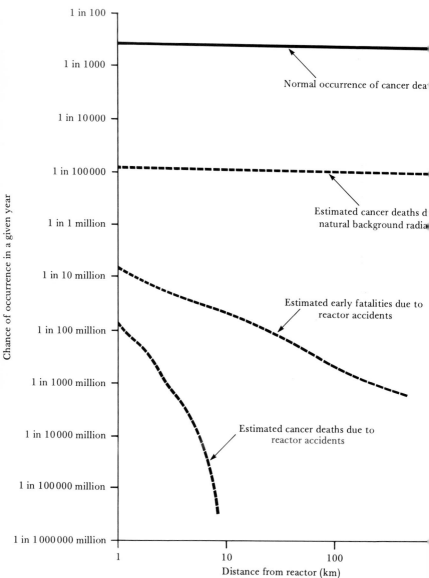

release of activity to the environment, because the majority of the activity was trapped within the primary containment. The radioactivity released to the environment was almost entirely the rare gases, krypton and xenon, due to a release of gases from the containment building during venting operations. Rare gases are not absorbed by the body: the more dangerous fission products are those like radioactive iodine, which can concentrate in the thyroid gland. Only 16 curies of iodine (I^{131}) were released. The average radiation dose to a person living within ten miles of the site was estimated to be $8/1000$ of a rem, about 7% of the annual background dose in the area. The health consequences of this release will be minimal: the Commission appointed by the US President reported that the number of possible cancer cases due to such a small release might be between zero and five over the next 30 years, compared with the 540000 cases that will develop in that population over the same period from other causes.[11]

Few accidents in industrial history can have been so comprehensively investigated and by so many committees. The most serious consequence of this accident has been the loss of public confidence in the US in the safety of reactors and in the competence of operators and of the regulatory authorities. In addition, the owners of the plant have sustained a serious financial loss. Two further points are worth noting. The first is the major discrepancy between the calculated health consequences, which will be minimal, and the perceptions of the risk shown by many local inhabitants (and the media), which caused much mental distress, a matter to which we shall return later. The second is that the very small escape of fission products – and, in particular, the very low ratio of iodine escape compared with the rare gases, krypton and xenon – has prompted a re-examination of the assumptions made in determining the consequences of reactor accidents involving the release of volatile fission products from over-heated fuel. It has now been realised that the actual releases of gaseous fission products and of particulate matter (apart from the rare gases like krypton) may be greatly reduced not only by engineered barriers as such but by natural physical and chemical phenomena, such as the formation of stable non-volatile chemical compounds in the presence of water, absorption on surfaces within the plant and the formation of stable aerosols which precipitate quickly in the near vicinity of a point of release and do not get widely dispersed. Very little of the activity may leave the site and the magnitude of the health effects predicted may therefore have been over-estimated. The assessment of actual releases (the so-called 'source terms') will play an important part in the next phase of the evolution of this subject. The magnitude of risk to individuals could be significantly reduced from those given in Figures 3.3 and 3.4 when due account is taken of the over-estimation of the source terms; this effect has been discussed in the calculations made for the Sizewell PWR case, but the revised estimates still have to be substantiated.

One of the major responsibilities of any management is to learn from accidents and mistakes. The detailed analysis of what went wrong at Three Mile Island and the recommendations for improvement made by the Kemeny Commission and many other investigations have certainly resulted in a number of changes. Improved procedures have been introduced in operator training in the US, in the layout of control rooms, in improved communications, better emergency control and in the analysis and recording of mishaps and malfunction of reactor components. The probability of the same type of accident happening again should have been sharply reduced as a result: Three Mile Island is therefore an example of nuclear technology 'advancing on the basis of a failure'.

Other major hazards

Probabilistic risk assessment techniques of the type described above have been applied to other industries besides nuclear power, as is illustrated in the recent Royal Society publication on '*Risk*'.[1] One of the best published examples is the Health and Safety Executive's study of the Canvey Island chemical and petrochemical complex on the Thames estuary. This complex consists of petroleum refineries and storage tanks for liquified natural gas, crude oil and other petroleum products, liquid ammonia and ammonium nitrate, hydrofluoric and other acids and other chemicals. Two Reports have been issued, in 1978 and 1981.[12, 13] The first Report indicated chances of up to one in 500 per year of accidents involving more than a thousand casualties, corresponding to a highest individual risk of one in 400 per year, but indicated also how these figures could be significantly lowered by improvements in safety arrangements. The second Report shows that remedial actions have been taken, and so effectively that the calculated risks have been reduced well below the levels indicated as possible in the 1978 Report. Some comparative results, taken from the second Report, are shown in Table 3.1. The calculated probabilities, particularly of large accidents, have been reduced by factors of 30 or more, with comparable reductions in estimates of the highest individual risk. It is a good example of

Table 3.1. *Estimated accident probabilities in the Canvey Island and Rasmussen studies*

Number of casualties exceeding	10	1500	12000
Estimated annual probability			
Canvey: 1978 estimates	1 in 210	1 in 350	1 in 3700
Canvey: 1981 estimates	1 in 1000	1 in 5000	1 in 100000
Rasmussen study: early fatalities	1 in 2 million	1 in 100 million	
Rasmussen study: fatal cancers	1 in 100000	1 in 1000 million	

the value of carrying out a comprehensive risk assessment as an aid to safety management. The revised, lower estimates still give estimated accident rates well above those for nuclear power plants, as is illustrated by the last two lines of the table.

Safety criteria and acceptability

In parallel with the development of the engineering discipline of risk assessment, ideas have evolved on acceptable criteria by which the safety of proposed installations may be judged. Two lines of thought can be distinguished: the protection of the individual and the need to avoid any large-scale disaster. Accidents involving multiple fatalities, particularly if these are concentrated in one area, with consequent local disruption, are regarded with particular aversion, even though they are rare and will only make a small contribution to the total overall mortality statistics. However, perhaps because of intensive coverage by newspapers, radio and television, these accidents create considerably more concern than do the steadily occurring accidents to individuals. For example, an aircrash killing 250 people in one day attracts far more attention than the road traffic accidents in the UK over a period of two weeks, although the same number of people are killed. In fact, the largest disasters to which mankind has been exposed are due to natural causes: hundreds of thousands have died in floods and in earthquakes. Large-scale disasters due to the failure of some human installation or artefact tend to be smaller in scale – hundreds of people can be killed in aircraft crashes, or thousands in disasters at sea. The phrase 'major hazards' actually began to be used in the 1970s and in the UK the Major Hazards Committee was set up following the accident in a chemical plant at Flixborough in which 28 workers were killed. There were 47 incidents in the British Isles between

1946 and 1975 in which 20 or more individuals were killed from man-made causes, giving an annual average death toll for this scale of incident of 69;[14] some statistics for the US are incorporated in Figure 3.2.

Comparisons with the frequency of other accidents, as a means of determining an acceptable level of safety, were suggested by F.R. Farmer in the 1950s and, in his pioneering paper on 'Siting Criteria for Nuclear Reactors' published in 1967,[3] Farmer suggested that an accident causing a release of several thousand curies of radioactive iodine (I^{131}) should not occur more frequently than once in a thousand reactor years, and that larger releases should be relatively less probable. This is a strict criterion since a release of 1000 curies of I^{131} should not have very serious consequences; no-one should suffer any immediate effects. Farmer calculated the consequences assuming a site with a population of four million within a radius of ten miles and showed that, for average weather conditions, a single release of 1000 curies of I^{131} might cause three additional deaths in the population (of four million) over a period of ten years or so after the accident. As we have seen, the Three Mile Island accident led to a far smaller release. The only reactor accident of this size, which has actually occurred, was to an early reactor of a type not operated today, at Windscale in 1957, when a temperature excursion led to a fire affecting several fuel channels and the consequent release of 20000 curies of I^{131} and other fission products of less radiological significance. A milk restriction area of about 200 square miles was defined two days later and finally lifted after six weeks. The accident was the subject of exhaustive enquiries. Some recent calculations[15] – based on the estimates of the medical consequences of exposures to low doses of radiation which were described in Chapter 1 – have concluded that 10–33 people might die as a result, over several decades and in an exposed population of about 20 million, which would be compatible with Farmer's calculation. As would be expected, there is no direct evidence of any such result.

The Advisory Committee on Reactor Safeguards (ACRS) in the US has recently published a comprehensive review of the approaches that have been made to the question of setting quantitative safety goals for nuclear power plants.[16] They propose limits to the frequency of occurrence of large accidents, such that there would be only a one in 100 chance, in a reactor life-time, of an accident causing significant damage to the reactor core. Such damage would not of itself result in any health hazard unless a large number of additional failures occurred to cause a release of radioactivity to the environment. Large releases should occur less than once in a million reactor years, with an upper limit of acceptability of five times that figure. The 'one in a million year' accident corresponds with a probability of early death, for an individual living near a reactor, of one in a million over his or her life-time, and a probability of one in 200 000 of delayed death due to cancer. The Committee points out that the application of its criteria would result in individual risks below those estimated to result from pollution emitted by comparable coal-fired power stations. Their target risk for multiple-fatality accidents is well below the estimated risks associated with an industrial complex of petrochemical plants such as Canvey.

All commercial reactors in Britain have to be licensed by the Nuclear Installations Inspectorate (NII) which was set up in 1959 and which now forms part of the Health and Safety Executive. The Inspectorate has issued[17] assessment reference levels for reactor reliability, which are essentially guidance to its own assessors on levels of safety at which a search for a further reduction of risk may not be necessary; these are not quantitative safety goals since there is still a requirement to seek lower levels of risk if cost-effective means of doing so are available. The overall acceptability of the safety case for a reactor is judged on the basis that all reasonable efforts should be made to reduce the probability of accidents, but the assessment reference levels give guidance on the acceptability of the measures used to limit or prevent specific fault sequences.

These reference levels are centred around the dose limits and Emergency Reference Levels (ERL) which were discussed earlier. Thus the dose to the public close to the site, from a single fault sequence, should be not greater than $\frac{1}{30}$ of the public annual dose limit of 0.5 rem in a reactor lifetime of 30 years and not greater than 10 rem (the ERL at which evacuation should be considered) in a reactor programme of 3000 reactor years (that is, 100 reactors, working for 30 years). The Inspectorate also require that fault sequences which might result in larger doses should be prevented by safety barriers or safety systems. If it is shown that an initial event can cause an accident giving more than 10 rem at the site boundary, with a frequency of up to once in 1000–10000 years, one barrier must be present in the design to prevent the release of activity; if such an accident could occur with a frequency of more than once in 1000–10000 years, at least two independent barriers should be used. Each barrier alone must be capable of reducing the dose at the site boundary to less than 10 rem. These rules correspond to a risk to an individual of contracting a fatal cancer of about one in 10 million per reactor year, with even lower risks of suffering from the effects of large-scale accidents.

While the Inspectorate has the duty of licensing a station, the responsibility for the safety of the plant, the staff and the public rests squarely on the reactor operator – in England and Wales, on the Central Electricity Generating Board. The CEGB has issued design safety criteria for its nuclear reactors, which are, of course, compatible with the guidelines published by the Inspectorate.[18] In particular, the CEGB defines targets such that the predicted accident frequency giving rise to doses of an ERL of 10 rem should not exceed one in 10 000 reactor years, and that the cumulative frequency of accidents leading to larger releases should be less than one in one million reactor years. These criteria are used in conjunction with numerical reliability analysis. The method of assessment was first applied by the CEGB and its design contractors to the Steam Generating Heavy Water Reactor Project, subsequently cancelled, then to the Heysham 2 AGR and later to the Sizewell B PWR. The method is being progressively developed and refined and is proving extremely useful as a means of obtaining a systematic approach in safety analysis. It is an aid which supplements, but does not replace, the long-established deterministic approach relying on scientific and engineering judgement.

It should be stressed that, in considering very improbable events such as major reactor accidents, the conscientious application of probability analysis and comparison of the results with some predetermined criteria can never be a sufficient proof of safety. One of the major findings of the Kemeny Commission on the TMI accident was that regulations alone cannot guarantee safety and that: 'It is an absorbing concern with safety that will bring about safety, not just the meeting of narrowly prescribed and complex regulations.' In the UK, the NII and the generating boards have always held wisely to the view that it is more satisfactory to ask the operating company to produce a comprehensive safety case that would stand up to rigorous assessment by a competent Inspectorate than to require them to meet a detailed set of regulations and criteria. A complete evaluation of safety includes consideration of operator training, maintenance schedules, safety review procedures and the rigorous control of modifications. The almost qualitative approach taken by the NII to major accident criteria is understandable. Assurance of reliability must depend on good practice, coupled with inspection and maintenance. Dunster has rightly argued in a recent lecture that progress towards quantitative safety goals should be deliberate rather than rapid.[19]

Perceptions of risks

It seems, then, that the criteria proposed for the safety of nuclear installations – like the results of the risk assessments – are well on the conservative side compared with risks normally accepted, or encountered, from other industries. But that is not how they are perceived and we now turn to the study of risk perception. This is another branch of the subject which is developing rapidly, to judge by the number of conferences arranged to discuss it, and it is again a relatively recent development, at least in its modern guise and formulation. Risk *assessment* is a branch of engineering and involves an attempt to arrive at a technical consensus independent of individual preference; risk *perception* involves the study of people's opinions and has its roots in applied psychology and sociology. The distinction should not be drawn too rigidly; practitioners of risk assessment exercise subjective judgement of technical issues, as do most, if not all, engineers, and the study of the perceptions of risks involves techniques of objective measurement.

The first approach is essentially that of the questionnaire. Selected groups of people are asked to rank various activities in order of perceived risk, or otherwise, to make comparative judgements of the risk. Such studies are always selective, in that they apply only to the selected groups, and the setting and interpretation of questionnaires is also a highly skilled business; the manner in which questions are asked can influence the response. Generalisations are more than usually risky in such a field, but it does appear that the perceptions or evaluations of risk held by some sections of the public differ very markedly from the technical assessments that have been made, and also from those based on actual mortality statistics. Typically, rare causes of death are over-estimated and common causes under-estimated. Accidents are seen as more important than diseases while, in fact, disease takes 15 times as many lives. Deaths from accidents, natural disasters, fires and homicides tend to be over-estimated – probably correlated with the fact that these are news, and well reported, whereas more ordinary causes of death are not, particularly those which claim only one victim at a time.

One example of the sort of results from such surveys is that given by Fischhoff & Slovic, and colleagues, carried out on student populations in the University of Oregon, and members of the League of Women Voters.[20] They were asked to quantify the risks of death from 30 different activities and technologies. Their gradings were compared with those of a group of risk analysis experts and for many of the activities it was possible to derive actuarial evidence of the average risks of death involved and therefore to compare risk perception with an objective standard. There seems to be a poor correlation between real and perceived risks, some of the estimates being in the wrong order, and the relative magnitudes of the different risks were not well perceived. Apparently, the people in this study failed to think about the size of the risks over a large enough range – the most risky perceived activity was thought to be only 16 times more probable than the least risky, while the corresponding real risks are spread over at least a factor of 1000, and probably 100 000.

In these and other studies,[21] nuclear power was rated as the most risky of all the technologies and activities listed – more risky than motor vehicles, aviation, smoking, etc. – by the groups of the lay public, but seen as the least risky of all by the expert group. Nuclear power was associated with a particular dimension of dread, of psychological pressure. This did not seem to be purely the fear of radiation effects, since X-rays were accepted as of little risk (less so by the expert group). This element of dread seems to have been due to the fear of unknown catastrophies, and to a quite exaggerated mental picture of a nuclear accident. In the studies referred to, more than 25% of the subjects expected accidents from nuclear power to involve 100 000 or more fatalities; more than 40% expected an accident involving 10 000 casualties if a disaster occurred in the year following the study! This mental image seems to be associated with the hazards of nuclear war; from the studies summarised in this chapter, a disaster of this magnitude originating from a nuclear power station is a most remote possibility, if credible at all. It is possible that the very discussion of low-probability hazards has led to the belief that they are likely, rather than to the realisation that they are most unlikely.

In addition to this 'exposed preference' method of assessing the perceptions of risk of some groups of the general public, there is also the 'revealed preference' method of study of what society is actually willing to pay, in terms of money and resources, for the avoidance of risks of different types. It is naive to think that society has arrived at an optimum balance between risks and benefits, but the comparison of money spent on safety in different activities is, in any case, revealing. It is repugnant to assign a money value to human life; no amount of money would compensate me for the loss of my wife's life, nor indeed of my own, but collectively we have to take such judgements; this is implicit in the distribution of money in the health service.

The cost of safety

It is a question of the use of resources: How much effort can be spared to extend one person's life? Table 3.2 shows some estimates of the money spent on saving one life, taken from tables compiled by Cohen[22] and by Kletz,[23] all approximately corrected to 1980 money values and converted to £-sterling by assuming £1 = \$1.6. All the figures quoted are for practices which are fairly common, so that the efficacy in terms of life-saving can be estimated, but which are not yet universally available or universally applied. The figures can only be indicative, but it seems safe to conclude that the spend on saving a life by many medical treatments is of the order of some thousands or tens of thousands of

Table 3.2. *Expenditure to save a life*

	£ per life saved*
Food for Third World starvation relief	7000
Medical	
Cervical smears	4000
Intensive care	15000
Heart isotope pacemaker	80000
Screening for lung cancer	80000
Accident prevention	
Traffic	200000
Smoke alarms in houses	300000
Steel industry	600000
Chemical industry	1000000

*1980 Money values, \$1.6 = £1

pounds. Kletz quoted a consultant (in 1972) querying whether it was ethical to concentrate resources costing as much as £5000 per case on the few patients in an intensive care unit! A much higher range of figures can be deduced from spends on industrial safety or on roads and traffic; society does seem to value accident prevention more highly than medical efforts to prolong life and spends hundreds of thousands of pounds per life saved; a value of $140 000 per life saved has been used explicitly in the US in making economic assessments of highway improvement measures. The chemical and pharmaceutical industries have been estimated as spending from hundreds of thousands to a few million pounds on industrial and environmental safety measures per life saved.

It is instructive to compare these figures with money spent to achieve higher degrees of radiological protection. The most cost-effective means of reducing the radiation burden borne by the general public would be to introduce measures to reduce the total exposure by X-rays during medical examinations, since the accumulated dose from this source is by far the largest man-made component of the average radiation dose, but the costs of doing so are not easily calculable. As far as industrial practices are concerned, the Nuclear Regulatory Commission has recommended that a sum of $1000 (£625) might be spent on reducing doses by 1 man-rem: since a dose of 1 rem to one man is associated with a risk of premature death of about one in 10 000, this is equivalent to the expenditure of £6 million per (statistical) life saved. We saw in chapter 2 that the NRPB has proposed values of the man-rem for doses to the public ranging from £20 to £500; at the higher doses, this corresponds to £5 million per life saved.

The actual money spent in the nuclear industry on radiological safety follows these guidelines. The average marginal cost of a major concrete biological shield is of the order of £100 per man-rem reduction in dose achieved, and plant modifications aimed at dose reduction fall in the range £1000–£2000 per man-rem avoided (£10–£20 million per life saved). A good example of spending to reduce the already very small radiation dose to the general public is a new plant to be constructed at Sellafield to reduce, principally, the discharges which are the cause of the highest exposures of the general public. The exposure of the normal fish-eating population in Cumbria and north Lancashire equals .01 rem per annum per individual. The critical group – the local fishing community – have in recent years accumulated 0.1 rem per annum, although discharges are reducing. The new plant should reduce these exposures by a factor of 5, at a cost of £50–£100 million. If the cost is averaged over the whole population, the effective value is at least £10 million per life saved, assuming that the linear dose–response relationship is applicable to the very low doses in question.

These figures relate to the reduction of small statistical risks, and may be compared with the costs of increased safety in other industries. The costs of reducing either the probability or the magnitude of serious accidents is another question again. Conceptually, this is a difficult question to answer. The basic design of a nuclear reactor is aimed at the safe control of nuclear power; no-one designs a reactor that will work and then adds features to achieve some level of safety. So one can look only at comparisons and at the cost of marginal improvements. Siddall[24] has argued that the increase above general inflation of nuclear power plant costs is largely due to the escalation of every aspect of regulatory intervention and the associated time-consuming procedures. He estimates the additional cost over the life-time of the reactor, and arrives at a figure of $188 million (£117 million) per statistical life saved, assuming a nominal 10% per annum interest on capital. Other studies have estimated that as much as one-half the capital costs of a nuclear power station built after 1978 are due to increased regulatory requirements introduced over the previous ten years.[25]

These enormous sums are truly being spent to reduce the margin of error in safety calculations and it is difficult to ascribe them definitely to a number of (statistical) lives possibly saved. Nor is life-saving the only consideration; as TMI has shown, a reactor accident is very costly and causes much distress, even if it is not in fact very dangerous, and there is therefore every incentive to achieve high reliability. Since the probabilities of large accidents are very small, however, the cost of achieving even higher standards of safety must be relatively very high in terms of the benefit gained. O'Donnell[25] has attempted some calculation of the cost–benefit ratio of improved safety measures, which range from £6 million per life saved for some effluent treatment systems to £1800 million per life saved for devices fitted to reduce the risk of hydrogen explosions in the containment building. It is certainly fair to conclude that very large resources are being applied to achieve very high levels of safety in the nuclear industry.

Siddall, in the paper to which we have referred, suggests that, if the highest degree of public safety is to be attained for a given total cost, all sources of money or other resources available for the improvement of safety should be pooled. This is not possible in the system we have today; the cost for each individual activity – nuclear power, coal mines, the health service – is financed out of separate allocations. In fact, we can conclude that the broad ratios between the monies spent on safety are more in line with people's perception of risk than with the quantitative technical estimation of them, made with the techniques we have today. This may not be surprising; in democratic societies, at least, we might expect the allocation of resources to follow roughly the popular sense of priorities. And although the question of how much to spend on safety arises acutely in the case of the nuclear industry, the dilemma is a general one and other examples of heavy spends on safety measures could be quoted, from the chemical and pharmaceutical industries in particular.

Since the resources devoted to achieving the levels of safety thought to be desirable are now so large, we have to ask why there is such a significant divergence between the technical assessments of risks and people's perception of them. We must beware of making an over-simple distinction between the quantitative assessment of risk-of-death, considering that as rational and right, and the public concerns, considering those as irrational and wrong. Lord Ashby, some years ago,[26] and several speakers in the 1981 Royal Society Symposium, have warned us that the adverse effects of hazardous events can only be fully evaluated in terms of human values and emotions, and that these are not easily quantified. There are two explanations of the divergence of views which have been noted. The first is that the public is not well informed about risks, and the second is that the techniques used in the assessment of risk are too narrow because they do not take account of the whole range of human concern.

As to the first, one must question whether enough is done to refine the information on the costs of safety measures and then to inform the public of the results. Is it truly the popular wish that resources devoted to safety in the nuclear and in other industries, relative to probable benefit, should be ten times those devoted to road safety and a hundred times those spend on improving medical care? It must be admitted that, in the case of the nuclear power industry, the record of programmes of public education is not hopeful;

people's attitudes are not easily changed once they are formed. But the record also shows that familiarity breeds trust: the level of concern about nuclear developments is markedly less in communities close to the plants than in those more distant from them, although it is the population close to the plants that face the highest risks. But, however difficult the task, the provision of accurate and adequate information is an inescapable respondibility of the industry.

Secondly, it is certainly the case that numerical risk-of-death is not a complete measure of detriment. People express various other reasons for their opposition to the nuclear power programme – dislike of centralised authorities, fear of the proliferation of nuclear weapons, and so on, and it is important to know whether or not the perceptions of risk associated with its development are compounded of these completely different considerations. Technologists must carry part of the responsibility for finding out what actually worries people, and for answering their anxieties as honestly as possible. It is anomalous and absurd that educated people should be fearful of accidents causing 10 000 or 100 000 casualties arising from nuclear power stations[20] when the probability of such an event is so very low; it should be appreciated more widely that all risks can only be described by a probabilistic approach and that absolute safety can never be guaranteed.

The discipline of risk assessment, like that of the medical and biological effects of radiation, is a subject of considerable difficulty and sophistication. The safety documentation of a reactor proposal is formidable – the Sizewell B Pre-Construction Safety Report submitted by the CEGB runs to 13 volumes – and the detailed scrutiny which is a necessary part of the licensing process can only be carried out by a strong professional team such as exists in the Nuclear Installations Inspectorate. But the general methodology and main assumptions used can be developed and refined under the normal disciplines of publication and of peer review. As in the field of radiation protection, we again face the difficulties of the proper pursuit of technical matters which are the subject of much public attention and emotion and of intense media interest. The Lewis Committee on Reactor Safety Assessment, which reviewed the Rasmussen Report, attempted to submit their findings, in draft, to a wide process of peer review and external criticism. They have this to say:

> Our interaction with the community has revealed two sour notes which need to be mentioned. The term peer means equal, and effective peer reviews are the only method known to the technical community for quality assurance in its product. However, in the arena of reactor safety, a peer comment has come to mean anything written by anybody to anybody, asserting anything about anything. The comment need not contain a better analysis, demonstrated technical expertise in the subject, evidence of error or conformance to any of the normal standards by which the technical community assesses a peer comment. Thus, responsible peer comments which do exist are immersed in a sea of others and this degrades the peer review process.

In other words, proper discipline is required here as in every other serious subject; good science or engineering calculation may not be enough to meet all the anxieties of the public, but bad science is positively dangerous. And it does not advance the cause of safety to inculcate exaggerated fears on the basis of undigested facts or of inadequate research.

4

Care for man's environment

The earlier chapters have been mainly concerned with the responsibilities of an advanced technological industry to its workforce and the general public in its present-day operations, though concern for future generations, as well as for people now alive, has been a feature of radiological protection since the 1930s, with the recognition of possible hereditary damage. Similar concerns have emerged relatively recently in other industries with, for example, a growing appreciation of the mutagenic properties of some chemicals. Looking to the distant future, another type of problem has attracted much comment in the scientific and technical press and occasionally in the national press as well. This concerns the possibility of causing *irreversible* changes in the natural economy of the planet through man's intervention. The historical example of such changes has been the destruction of soil fertility over large areas, due to over-grazing, over-cultivation and the destruction of plant and tree cover. This process is still going on in some parts of the world in which wood is a primary commercial product and in others in which wood is the only fuel that the local population can afford: the destruction of forests may emerge as one of the worst ecological problems later in this century. The increased burning of fossil fuels may be a contributory factor to two other effects which are receiving increased attention – the destruction of plant life due to 'acid rain' and the possibility of climatic changes due to the rising proportion of carbon dioxide in the atmosphere. Another example is the possible consequences of the injection of higher concentrations of fluorocarbons into the upper atmosphere on the intensity of the ultraviolet radiation reaching the earth's surface. Typically, the evaluation of questions like these involve very difficult and uncertain extrapolations of complex scientific data. For instance, it is known that the proportion of carbon dioxide in the atmosphere has increased by 10% since 1900. The reasons for this are not clear, though the increased burning of fossil fuels is an obvious source, and there is room for much speculation about the possible consequences if this trend continued. However, it would at least be prudent to evaluate and monitor changes carefully before unwelcome and irreversible effects are actually upon us.

In the case of the nuclear industry, concern about the effects on the environment in the distant future has been concentrated on the long-term management of the radioactive wastes which are necessarily produced as a byproduct of the operation of nuclear reactors. Judging by some articles and statements, one might imagine that the importance of this subject has only recently been realised and that its dimensions are so horrendous that the whole nuclear industry should be strangled for this reason alone. Historically, it is not a new subject at all: the disposal of radioactive wastes was seen as an essential part of a large-scale nuclear energy programme from the start, as was envisaged by Sir John Cockroft in his Joule Memorial Lecture in 1951, five years before the opening of Calder Hall. He said:

> The disposal of radioactive wastes at one time seemed to be one of the major problems of a large-scale atomic energy programme . . . The fission products consist of about 30 radioactive elements. The activity of some of these decays fairly rapidly. The problem is therefore to separate by radiochemical methods the fission products which are long lived and to concentrate and store them.

Methods are now being developed which should make this feasible.

Until the late 1950s, disposal of radioactive waste was subject only to the law governing ordinary waste disposals and there were no specific restrictions on the keeping or use of radioactive materials, except on the premises of the AEA and other licensed sites. The first comprehensive set of principles was contained in a White Paper published in 1959[1] and regulations were embodied in the Radioactive Substances Act of 1960. Mounting concern for environmental pollution of all types led to a number of other major acts in that period starting with the Clean Air Act of 1956, and ending with the Health and Safety at Work Act of 1974 and the Control of Pollution Act 1974. The environmental departments set up an expert group to review the principles of radioactive waste management, which reported in September 1979,[2] and this Report contains an exhaustive study of the working of the various acts over the last 20 years. The latest statement of government policy is a White Paper published in 1982.[3] The responsibility for the safe management of radioactive wastes and for making proposals for their ultimate disposal rests with the nuclear industry, but the responsibility for defining the regulations that must be obeyed, and for conducting the necessary research on which the regulations can be based, rests with the environmental departments, led by the Department of the Environment. The departments are advised in this task by the Radioactive Waste Management Advisory Committee, an independent body.

We have another problem here, which, in principle, is not unique to the nuclear industry. Any time that material is dug out of the earth at one point, transported somewhere else, transformed into materials that do not occur naturally and eventually disposed of after use, the environment of man is altered to some extent. The control of many types of waste, not only radioactive wastes, has regional, national and, in some cases, world-wide implications. It is perhaps ironic that so much attention has been focussed on the disposal of a type of waste, radioactive waste, where the hazard demonstrably decreases with time, as compared with other types of potentially toxic substances, mercury, arsenic, cadmium, lead, asbestos, etc., which have an infinite life. There probably exists today a larger body of knowledge and ability to quantify the problem in the case of radioactive waste than is the case with many other toxic substances, and perhaps the very existence of this knowledge has helped to focus public and political attention.

Types and quantity of radioactive wastes

The term 'radioactive waste' is itself something of a misnomer. What we are concerned with are different categories of waste material of which some of the constituents are radioactive. Such waste materials arise at every point of the nuclear fuel cycle, from digging uranium out of the ground, refining it and enriching the resulting pure uranium in the fissile isotope U^{235}, forming fuel elements, irradiating these in a nuclear reactor and eventually recovering the spent fuel, and, in the UK, processing it chemically to separate out the materials which might be used as future fuels, uranium and plutonium, from the fission products. Characteristic constituents of the various waste streams are, then, as follows:

(i)   Uranium mill tailings containing all the radioactive decay products in the naturally radioactive series resulting from the decay of the uranium in the ore. Since at radioactive equilibrium there are 19 members of this series, the greater part of the radioactivity stays behind in the tailings and this includes some quite long-lived components, such as radium (Ra^{226});

(ii) The fission products that are necessarily formed during the irradiation of the uranium fuel itself. They consist of a mixture of radioactive isotopes of a large number of chemical elements, most having short or very short half-lives and a few with long half-lives;

(iii) The actinides, that is, heavy elements resulting from absorption by uranium nuclei of some of the neutrons produced in the fission process; many are α-emitters, and comparatively long-lived. The most important actinide is plutonium, which is stored and not treated as a waste material, though small amounts appear in other waste streams;

(iv) Radioactive materials formed by interactions between neutrons and the structural materials of the reactor.

Large quantities of waste material are produced at uranium mines; these wastes contain some unextracted uranium and all the radioactive decay products originally present in the ore. All that has been done is to move some material from one part of the earth's crust to another. However, the fact that the material has been brought to the surface and finely ground increases the possibility of its being dispersed, and reaching man. Further, the radioactive gas radon, a decay product of uranium, is able to escape more easily from the disturbed material. The ideal procedure would be to return the wastes to the original mines but this is seldom economic. In practice, the wastes need to be stabilised, generally by earth cover and re-vegetation, to prevent wind and water erosion and to ensure that the public health hazard is kept at a very low level. No uranium mining is carried out in the UK and the subject will not be considered further here.

Radioactive waste streams are commonly characterised

as low, intermediate or high level, according to their specific activity; that is, radioactivity per unit weight or volume. In general, the term 'low-level waste' is used to describe those wastes that can be disposed of directly to the environment under present regulations, whereas the intermediate-level wastes require some form of treatment before disposal, and high-level wastes must be isolated. The development of the principles governing our present practices of disposal of low-level wastes in this country was discussed in Chapter 2 in which we saw that the application of these principles to the discharge to the environment of liquids and gases with low levels of activity results in only very low exposure to radiation. Two categories of solid wastes are presently disposed of beyond those of such small activity, such as old luminous dials and some hospital wastes, that they can be consigned to municipal dumps. Low-level solid waste consisting principally of contaminated laboratory equipment, protective clothing, and miscellaneous rubbish is disposed of at the Drigg Site; a fenced site of some 300 acres, four miles from Sellafield. The waste is placed in trenches which are isolated from the underlying sandstone by a layer of clay and covered by at least one metre of soil. The trenches drain into a tidal stream. Wastes of somewhat higher specific activity are disposed of at sea under a treaty known as the London Convention; the technical considerations which led to the definitions of acceptable waste levels for disposal at sea are described below.

All other categories of wastes are currently stored on the sites at which they arise. The approximate volumes and specific activities of the wastes arising up to the year 2000 are given in Table 4.1.[4] The table is arranged in order of decreasing specific activity, and it is immediately apparent that only small volumes of high-level wastes exist, though they contain 95% of all the radioactivity, while comparatively large volumes of low-level waste trash are generated. In general, the volume of wastes increase as the specific activity decreases. The treatment and means of storage and disposal appropriate to each category of waste are determined by the specific activity and the half-lives of the radioactive components; that is, by the rate at which the activity decays away. The same principles apply in each case: no radioactive material is discharged into the environment unless and until the activity is low enough and the local dilution large enough to reduce any exposure to acceptable levels. It has been accepted that disposal is preferable to indefinite storage; 'disposal' signifies no need for continued surveillance and no intention to retrieve, though, of course, radiological monitoring can continue. So all categories of waste have to be put into a form in which safe disposal can be envisaged, though a period of storage is often necessary and isactually beneficial in the case of the high-level wastes, as discussed below. It will

Table 4.1. *Estimated radioactive waste arisings in the UK. Volumes to the year 2000 after treatment*

	Volume of treated waste (m^3)	Specific activity (curie per m^3)	
		α	$\beta-\gamma$
Highly active waste from re-processing after vitrification	940	17 000	4 500 000
Wastes with high α-activity, e.g. fuel cladding; re-processing sludges	39 000	40	2000
Intermediate-level wastes from re-processing and laboratories	22 000	<1	50
Wastes from reactor operation	24 000	0.01	100
Low-level wastes*	490 000	<0.001	0.01

*As currently disposed of at the Drigg site.

be convenient to deal with the different categories in turn, starting with those that are currently discharged.

Treatment of low-level wastes

The low-level solid waste is slightly contaminated trash – like gloves, plastics, glassware and so on, that arises in every laboratory, plant, industry or hospital where radioactive sources and materials are used. These wastes are suitable for disposal in shallow trenches surrounded by impermeable material like clays which impede water movement. The Drigg Site is licensed to accept material with specific activities lower than 20 millicuries per m^3 +-activity and 60 millicuries per m^3 β-activity; a recent review found no reason to alter this regulation.[5] The total arisings of 20000–30000 m^3 per year amount to 50 thousand tons, which is small compared with the 15 million tons of coal ash or the five million tons of toxic chemical waste disposed of annually in this country. Even the total amounts of radioactive materials in this annual disposal of low-level waste are relatively low, at most, 150 curies of α- and 450 curies of β-activity. More radioactivity reaches the environment from other sources; for example, the 100 million tons of coal annually brought to the surface contain about 500 curies of α- and 600 curies of β-activity, and this amount of activity is subsequently dispersed during combustion or in the ash. The extra care that is taken of the radioactive wastes sent to Drigg does not arise because of the total amount of activity but because it is concentrated in one place and the principle of protection of a critical group – the local population – applies.

Wastes of a higher, but still low, specific activity can be disposed of in the deep ocean under the terms of the Convention on the Prevention of Marine Pollution by Dumping of Wastes, the 'London Convention', signed since 1972 by 47 countries. The limits on the amount of radioactive materials that can be so disposed were arrived at after a series of studies started by the International Atomic Energy Agency. In addition, the actual disposals carried out by European nations take place under a consultation and surveillance mechanism operated by the Nuclear Energy Agency of the OECD. Each dumping operation is the subject of a specific licence, issued in the UK by the Ministry of Agriculture, Fisheries and Food. The Convention currently limits annual disposal at one site to 100 000 tons of material of specific activity less than 1 curie per ton of α- and 100 curies per ton of β/γ-activity. The actual disposals are a small fraction of this limit. During the last few years, the UK has disposed of 2000–2500 tons of cemented waste each year, with total α-activity of up to 2000 curies and β/γ-activity up to 100 000 curies, mainly tritium. This operation of sea disposal has also recently been used by other countries including Belgium, Switzerland and the Netherlands.[6]

These activity levels are very small compared with the natural radioactivity of the sea. The North Atlantic, a volume of some 200 million cubic kilometres, contains about 500 million curies of α-activity, mainly due to the uranium content, and some 60 thousand million curies of β-activity, mainly due to potassium (K^{40}). The complete dispersal of a few thousand curies of α-active material and a few hundred thousand curies of β-activity would clearly add a negligible amount to what is already there. So the only question is whether some local mechanism of concentration might operate to short-circuit complete mixing and result in a critical group suffering an appreciable additional dose.

The solid waste is packaged in drums, in an inner container surrounded by concrete, so that, as a minimum, the packages reach the sea floor intact. The current dumping area is well away from the continental shelf, with a water depth of 4400 metres. The only realistic pathway back to man that could be envisaged is one where the radioactivity is dissolved in the water of the deep ocean and subsequently contaminates shallow waters used for fishing. The mathematical model of critical pathways to man, on which the London Convention limits are based, makes a number of conservative, even pessimistic, assumptions. The model envisages a disposal of 100000 tonnes of packaged waste every year for 40000 years; the amounts actually disposed are a few per cent of this quantity. It is assumed that the activity is immediately dissolved or dispersed when the containers reach the sea floor whereas, in fact, the drums are likely to retain their integrity long enough for some of the activity to decay before it is dispersed. In addition, the concentrations at the sea bottom are used as if they were at the surface in order to calculate possible concentrations of a radioactive species in fish, whereas, in practice, considerable dispersion is likely to occur. Monitoring of the North Atlantic site has not identified any increase of radioactivity beyond background levels. The use of this site was last reviewed by the Nuclear Energy Agency of the OECD in 1979; the conclusion was that the disposal operations could safely be continued or even increased and that no further review was necessary until 1984.[6]

At the time of writing, the continuation of this operation of disposal of low-level waste into the Atlantic deep is a matter of some controversy. A majority of those nations represented at a meeting of the London Convention in 1983 voted in favour of a moratorium for two years pending a scientific review of the operations. The resolution was not binding on member countries and the UK Government has signified its intention to proceed unless any fresh evidence comes to light indicating that the operation will cause any hazard. No scientific evidence relevant to the use of the North Atlantic site was laid before the 1983 meeting, and the work that has been done for the Nuclear Energy Agency since their last review in 1979 has not revealed any adverse features. Advice from the National Radiological Protection Board is that this category of waste should continue to be disposed in this way on grounds of radiological protection, a conclusion which was endorsed by the Radioactive Waste Management Advisory Committee, an independent body set up to advise government ministers, in its last annual Report.[7] It is to be hoped that this technical advice will prevail and that the UK can continue to use such disposal routes as are judged to be best for each category of waste.

Intermediate-level wastes

Many types of intermediate-level wastes referred to above and in Table 4.1 are too active for disposal by existing routes and divide into two broad classes. The first consists of wastes arising at reactor stations and includes ion-exchange materials and sludge arising from liquid clean-up, and some used reactor components; these contain little long-lived α-emitting material and will typically decay to very low levels in a few hundred years. Such wastes will be converted into a stable solid form by incorporation in cements, bitumen or polymer-impregnated cement with some pre-treatment where necessary. In a stable, immobile form they will be suitable for shallow burial, at about 20 metres, in trenches that are engineered to a higher standard than the ones at Drigg. These trenches would be lined and filled in with cement and packed with clay to retard water movement to and from the wastes, and to act as an absorbent barrier to any activity that might be leached out in time (Figure 4.1). In order to establish radiological safety, experimental determinations will be made of the activity that could be leached by groundwater and the rate at which it could be dispersed into the environment.

The second category of intermediate-level wastes are those which contain relatively high quantities of α-active species, some of which, for example some plutonium isotopes, will decay slowly, with a half-life of thousands of years. Such debris includes the metallic cladding from fuel elements, with particles of fuel clinging to it, sludges from some ponds in which fuel elements have been stored, and plutonium-contaminated materials from fuel fabrication and re-processing plant. Such wastes can be treated in two ways: either they can be subjected to vigorous de-contamination procedures which result in a concentrate to be added to the high-level waste, and a larger volume of low-level waste; or the wastes can be immobilised, after appropriate treatment, in cement or bitumen. In the latter case, the α-active blocks can easily be stored in any dry situation but will eventually have to be disposed underground, at depths up to 300 metres to guarantee containment and isolation for a sufficiently long

Fig. 4.1. Shallow repository for short-lived intermediate-level wastes.

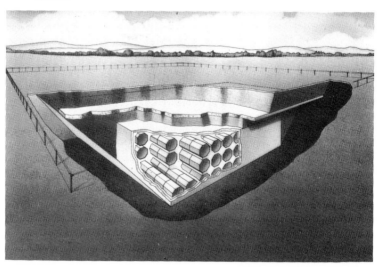

time (Figure 4.2). Pilot disposals of similar waste have been conducted in an abandoned salt mine in Germany, and other existing cavities could well be suitable, particularly if they are dry. Otherwise, a new excavation will be necessary in a

Fig. 4.2. Deep repository for long-lived intermediate-level wastes.

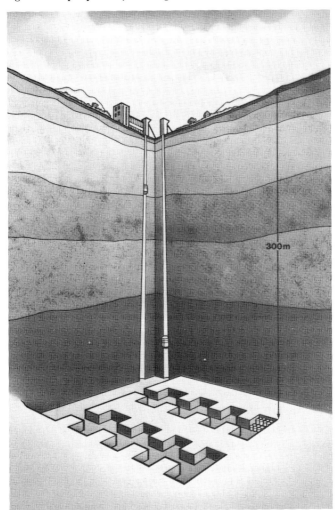

suitable geological environment. In either case the repository would need sufficient capacity to accommodate
25 000 m^3 by the year 2000.

As the recent White Paper[3] explains, there will be a
need before the end of the century to establish one or two disposal sites for intermediate-level waste in the UK, for both
shallow and deeper burial of these wastes. The radiological
criteria to be met will have to satisfy the environmental
departments, just as the actual operations on any new site
will have to be licensed by the Nuclear Installations Inspectorate. The principles to be followed in establishing the
safety of such a site are those that have already been
described: namely, the identification of all credible routes by
which the buried radioactivity might reach man; the identification of a critical group of the population in each case; and
the calculation of the maximum radiation dose to which such
a group may become exposed. The data required include the
rate at which activity might become dispersed, for example
by leaching into groundwater, and the rate at which the
active species could move in the local environment. Such
data can only finally be generated by comprehensive experimentation on specific sites, but generic studies by the NRPB,
using typical values of the parameters, have indicated that
repositories such as those shown in Figures 4.1 and 4.2
should be acceptable.[5]

High-level wastes

The main interest and concern for the future centres
upon the fate of these wastes, which contain about 95% of the
total activity and in the most concentrated form. These arise
in the spent fuel which is discharged from reactors and constitute the fission product and actinide stream, which is separated out after chemical reprocessing. As can be seen from
Table 4.1, the quantities of these materials are comparatively

small. The annual flow of material through a reactor designed to generate 1000 MW of electrical power results, in one year, in the production of something over one ton of fission products and a few kilograms of actinides. The heat generated by the radioactive decay of materials of this high specific activity is quite considerable; the heat emission falls as the activity decays. This is the reason why spent fuel rods are stored in cooling ponds after discharge from the reactor, and why the high-level waste, after chemical separation, still has to be cooled. High-level waste is currently stored in acid solution in cooled stainless-steel tanks at Sellafield, built behind thick concrete walls to shield the operators from radiation. Currently, 12 tanks contain the 1000 m^3 or so of highly active waste which has been generated by the entire nuclear programme in the UK to date, and a somewhat lesser volume of liquor of much lower activity is stored in tanks at Dounreay. The cooling systems consist of stainless-steel coils through which water flows (Figure 4.3). The tanks and cooling systems have proved durable and satisfactory but they need some supervision and would eventually have to be replaced.

Vitrification and storage

It was perceived in the early days that storage of highly active heat-emitting wastes in acid solution could only be regarded as an interim solution. Containment under controlled conditions would be simpler and cheaper if the wastes were converted to a solid form. Work on processes of solidification started at Harwell in the 1950s and papers for the second Geneva Conference in 1964 recorded the development of a vitrification process in which the elements of the waste are incorporated as constituents of a glass which can be manufactured by adaptation of fairly standard glass-making technology. Several other solidification techniques have been proposed and are mentioned later. But glass has remained the leading choice as a waste form at present and the technology has been brought to a commercial scale in France, where the plant at Marcoule has been transforming high-level waste into blocks of active glass since 1978 (Figure 4.4). British Nuclear Fuels Limited have announced that

Fig. 4.3. High-level waste storage tank, showing cooling coils being lowered into a tank under construction.

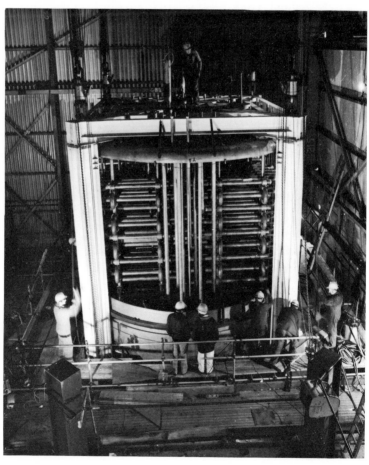

Fig. 4.4. AVM (Atelier Vitrification Marcoule) process for the vitrification of high-level wastes. In this continuous process, wastes are partly calcined in a rotary kiln. The calcine, a powdery substance, flows into a melting pot into which glass powder is also injected. The mixture is melted at 1150°C and poured into stainless steel containers which are then welded shut.

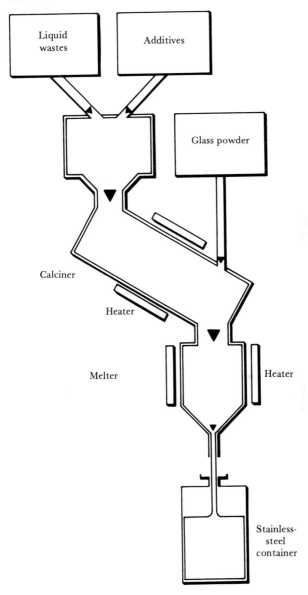

they intend to build a vitrification plant at Sellafield similar to the French plant at Marcoule and to bring it into operation by the end of this decade.

The blocks of waste glass will still be emitting a fair amount of heat so they must be stored under conditions where they can be cooled artificially for a further period to avoid any danger of melting or de-vitrification of that part of the glass which could reach high temperatures. In addition, since the blocks will still be highly radioactive, they have to be adequately shielded. These two conditions for safe storage can be met very simply. The glass blocks, in suitable metal canisters, can either be stored in a pool of water, similar to that in which fuel elements are stored, or they can be placed in a store surrounded by concrete shielding in which they can be cooled by a current of air. The French actually use air cooling in Marcoule (Figure 4.5). From the point of view of the ease of inspection and proof of safety, there is something to be said for an air-cooled store in which the blocks can be kept dry. Glass is an inert and durable material: so long as it is dry there is no credible mechanism by which the activity can disperse and get back to man. The conditions to be maintained are simple – merely the continuation of an air current and, no doubt, of monitoring of the exit air to confirm that the canisters containing the glass are intact. Access of the public to such a store would have to be controlled in their own interest. It seems a simple enough management operation to keep in being.

The need to consider a final stage of ultimate disposal has arisen because of the long-term nature of the potential hazard and the worry that the proper supervision of the wastes in stores could not be guaranteed for a sufficient length of time. In this country the most complete articulation of this point of view occurs in the Sixth Report of the Royal Commission on Environmental Pollution,[8] published in

Fig. 4.5. Air-cooled store for vitrified high-level wastes.

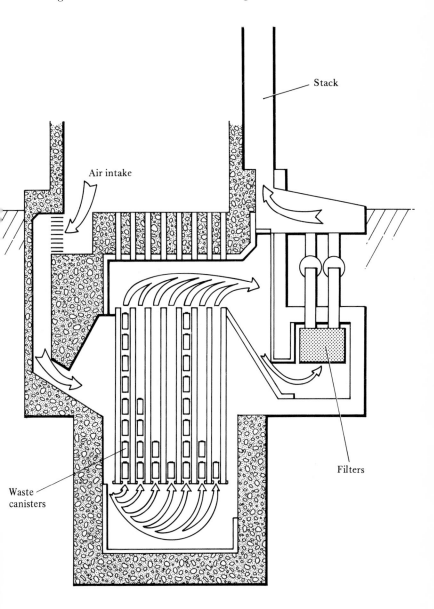

1976. The Commission accepted that some continuity must be assumed in human affairs and institutions and in the ability of future generations to maintain the containment, but their conclusion was:

> We are agreed that it would be irresponsible and morally wrong to commit future generations to the consequences of fission power on a massive scale unless it has been demonstrated beyond reasonable doubt that at least one method exists for the safe isolation of these wastes for the indefinite future. The use of this method might appear unjustifiably expensive and alternative approaches might be adopted in the event. But it would and should be available as a fall-back position if circumstances demand it.

The Government's response to the Royal Commission's Report was to expand the nuclear industries' research programme, and to transfer responsibility for the research from the Department of Energy to the Department of the Environment. Research was aimed at evaluating three options for final disposal – deep under land in suitable geological strata, or on the deep ocean bed, or underneath the deep ocean bed. It is evident that, because of the relatively high initial heat emission from the highly active wastes, there are operational and safety advantages in extending the period of cooled storage for several decades, and perhaps for 50 years or more; these considerations do not apply to other categories. In 50 years, activity levels and heat emission will have declined to about one-tenth of what they are three years after discharge from a reactor. Several benefits result from such an extended storage period. Firstly, it is an advantage to be able to inspect the waste packages during the period of maximum heat emission when the temperature gradient from the centre to the outside of a glass block is high, and to take remedial action if necessary. The effect of a period of storage on this temperature gradient is illustrated in Figure 4.6 which shows the temperature distribution across a typical glass block, 50 cm in diameter, containing 25% fission products, at various times of storage in a pond of water at 25°C. Secondly, all subsequent engineering operations are simplified once the glass blocks do not need to be artificially cooled. Thirdly, the effects of disposal on the surrounding media – geological strata or sea-bed sediments – are smaller and easier to calculate if the heat emission from the glass blocks is low before final disposal is implemented.

Options for final disposal

The operational need for disposal of high-level wastes will, then, not occur for some time. However, the need remains to prove options for disposal that do not make further demands for maintenance and inspection. To quote the recent White Paper: 'We in the present generation have a clear moral duty to formulate the options as we see them at present, and to develop the supporting scientific and technical knowledge so that a future generation will be better placed than we are to make the eventual choice.'[3] The White Paper goes on to state that the feasibility of geological disposal has been established in principle as a result of extensive research in many countries, and that research should continue in order to bring the knowledge of other disposal options up to the same level. Before reviewing the present state of knowledge, it is appropriate to set out the important technical facts.

The key characteristic of the waste from the point of view of adequate containment is the life-time of its constituents; Figure 4.7 shows the activity in curies for the various constituents in the waste.[9] For convenience, the quantity of waste is that produced by the generation of 1000 MW-year of electricity, the approximate annual output

Fig. 4.6. Temperature distribution in a 50 cm diameter block of vitrified high-level wastes stored in water at 25°C, at various times after manufacture.

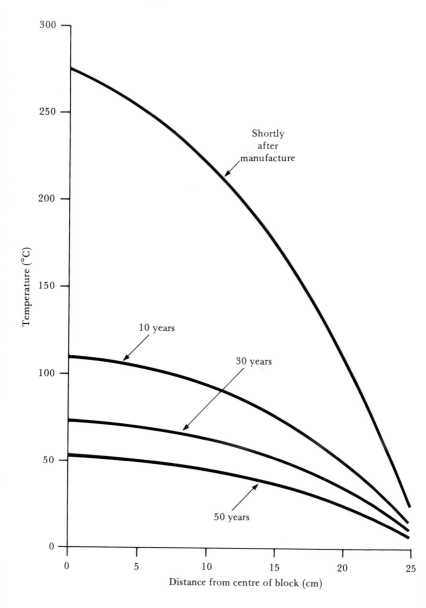

Fig. 4.7. Decay of total and individual activities of fission product wastes from the generation of 1 gigawatt year of electricity (1 GW(e)yr) in an AGR.

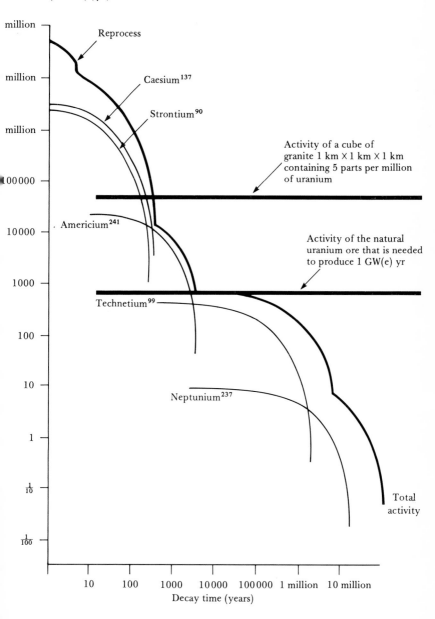

from a twin reactor AGR station; the waste from a PWR or any other reactor of similar output is very similar. Re-processing is assumed to occur after about four years of discharge of fuel from the reactor, with the expectation of removal at that stage of 99.9% of both the plutonium and uranium. The first important point is the overall rate of decay. The activity in the fuel at discharge from the reactor is about 1% of the total activity in the fuel rods at shutdown because of the very rapid decay of short-lived fission products. The activity decays by another factor of 10 in the first few years and, for the next 300 years, the activity is dominated by the decay of caesium and strontium isotopes which are β/γ-emitters. The activity has decreased by a factor of over 100 000 within a few hundred years after shutdown, after which the significant species are americium[241] and, later, technetium[99], a β-emitter of 210 000 years half-life, and neptunium[237], an α-emitter of two million years half-life. In order to provide a standard of comparison, the diagram includes the total activity of the quantity of natural uranium ore that has to be dug out of the ground and refined in order to fuel the reactor to generate 1000 MW for a year – about 150 tons of uranium for an AGR reactor. Included also in the diagram is the total activity due to the natural uranium content of a block of granite, a cube of 1 kilometre edge, containing five parts per million of uranium in equilibrium with all its radioactive products, which is a typical uranium content for granite. It follows immediately that the activity in the long-term component of this waste is a very small fraction indeed of the natural activity in our environment, and that this will be true for a nuclear power programme of any attainable size. Table 4.2 shows the estimated total uranium in the top kilometre of the earth's crust, the uranium content of the oceans and the upper limit of the known and estimated tonnage of economically extractable uranium ore. Since the

long-lived activity in the waste will be less than that of the
necessary ore, using all the ores in thermal reactors would in
the long run add a negligible component to the earth's
radioactivity. Indeed, even if all the estimated world reserves
of uranium were to be used in a fast reactor fuel cycle, with a
50-fold increase in the efficiency of using the uranium (an
enormous nuclear programme which could not be accom-
plished for centuries, if ever) the long-term component of the
resulting waste would be about $\frac{1}{100\,000}$ of the natural radioactivity in the earth's crust. Since it is not reasonable to
postulate very rapid mechanisms of uniform dispersion
throughout the world, particularly once the waste is solidified, we are therefore dealing with a question which is a local
and not a global concern. We must again ask: What are the
pathways back to man? What is the composition of the critical group likely be affected and by what criteria should we
seek to ensure their safety?

Strict application of the radiological protection principles that have been discussed in Chapter 2 to situations in
the distant future is not straightforward. Clearly, any
attempt to define a critical group in the future can only be a
hypothesis dependent upon assumptions about the future
distribution of population and its life-style. Attempts to calculate the total doses to populations are more difficult and
the uncertainty is greater the further into the future the calculation is carried out. The simplest assumption would be to
apply the same dose limits to future populations as to present
ones, and to aim to restrain doses to individuals likely to be

Table 4.2. *Estimated quantities of uranium*

Uranium in crust (1 km)	4 000 000 million tonnes
Uranium in oceans	4000 million tonnes
Uranium in useful ores (upper limit)	5 million tonnes

most at risk – for example, from drinking water from supplies close to a repository – within the present rules affecting critical groups. The application of an optimisation procedure – to try to calculate what is 'as low as reasonably achievable' – will be more difficult since large values of collective dose can result from adding together very small doses over very long time periods and large populations; these would really have no significance because of the much larger exposure to background radiation. One way of setting a sensible limit to such calculations is to agree on the value of an annual dose which is insignificant, in a manner similar to that which has already been proposed. Another approach, which has the same effect as setting a limit, is to introduce some method of discounting the cost of future detriment, on the usual economic argument that a misapplication of resources results unless future costs are discounted to present-day values. Such calculations have a useful illustrative value, but there will be great difficulty in deciding on a reasonable discount rate; this type of problem is not a typical one of economics. It might be more practicable to agree a time which would be a reasonable limit for the calculation of future dose since the radioactivity will be decreasing. On the basis of the activity curve (Figure 4.2), a time of a few thousand years sets a reasonable common-sense limit, after which the hazards resulting from waste disposal are surely small compared with many other natural hazards. However, these arguments about time limitation have yet to be developed rigorously; for the present, attempts will continue to calculate individual doses that could arise in the future, without any time limit.

Geological disposal of high-level waste

The application of any criteria of dose limitation to events so far in the future demands the development of predictive models that can be verified by experimentation in a reasonable time. A very large effort is being expended on their development, particularly into the options of geological disposal, in the US, Canada, Japan, Sweden, and the EEC countries including the UK. In a geologically stable region, the only conceivable mechanisms by which the radioactive species buried underground could get back to man is through interaction with groundwater, and the first defence against this possibility is the properties of the containment immediately surrounding the solidified waste itself – the engineered barriers. The principal long-term barriers are the properties of the host rocks themselves (Figure 4.8). They should be stable, the water flow through them should be slow, and the absorbent capacity of the rock itself for the radioactive species should be high.

The engineered barriers are the resistance to any leaching afforded by the properties of the vitrified waste itself, of the container in which the glass block is enclosed and of any packing or in-fill material surrounding it. As a result of extensive tests, there is now considerable confidence that the glasses that have been developed in many countries can withstand both the heat and the radiation that is emitted as the wastes decay, without serious degradation. A number of corrosion-resistant materials are available for the first container immediately surrounding the glass – stainless-steel with over-cladding of lead, cast iron, titanium, copper, have all been evaluated; our own work points to a choice between a very resistant titanium alloy or a thick container of cast iron. Outer packing materials, such as clays with very favourable absorptive properties, will delay both the ingress of any groundwater from the surrounding strata and the egress of any leached wastes. The leach rates of the glasses themselves are low at those temperatures at which contact with groundwater may take place. All these studies point to the conclusion that, given suitable repository designs, the

Fig. 4.8. Deep underground disposal of vitrified high-level wastes.

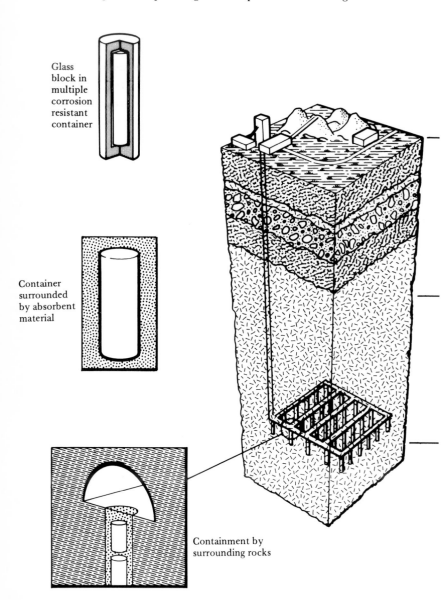

Glass block in multiple corrosion resistant container

Container surrounded by absorbent material

Containment by surrounding rocks

engineered barriers can be expected to retain their integrity for at least many centuries, and perhaps for very much longer, with slow release of the activity thereafter.

To estimate the efficacy of the various barriers in practice, it is necessary to set up a mathematical model which relates the rate at which various species appear in the biosphere to the various parameters characteristic of the engineered barriers and of the geological containment itself. It is then necessary to make some assumptions on the mechanism by which any radioactivity reaching the biosphere will actually reach a critical group of the population. A mathematical model of this sort has been set up by the National Radiological Protection Board and used by them to determine the sensitivity of the results to variations in the values of the different parameters.[10, 11] Used in this way a model is a valuable guide to the most sensitive areas for future research. The NRPB authors have assumed that groundwaters flowing through a repository discharge into a freshwater body used by man, at the end of its flow path, and that the critical group is therefore the people drinking this water. More recently[12] they have considered coastal sites where dilution in the sea might result in even lower maximum dose. A few conclusions are as follows.

The engineered barriers should prevent any attack on the vitrified waste by groundwater for a period of some centuries at least, during which no escape can occur. At longer times, many thousands of years, the highest radiation doses (though still low) are likely to arise from only a small number of long lived isotopes, the most significant of which is neptunium^{237}, one of the actinide elements. The NRPB authors comment that further data on the metabolism of neptunium are required since the currently recommended values are based on the results of a limited number of experiments, which may over-state its metabolic effects. The NRPB study also showed the importance of geological and hydro-geological parameters, such as groundwater velocity, and the sorption properties of the rocks for the neptunium species.

Absolute values of doses to individuals can only be derived from such models when these geological and hydro-geological parameters are known. Models of this sort were applied in a comprehensive manner by the Swedish KBS project under the imperative of meeting conditions of a 'Stipulation Act' under which reactor operation would only be allowed if the operators could show that the resulting wastes could be finally stored with 'absolute safety'. Much argument followed on a proper definition of 'absolute safety'. The 'KBS project' was formed to complete, in one year, the technical design for an ultimate waste disposal system in Sweden, to establish criteria for design and geological characteristics and to develop the safety methodology. Some 300 or more researchers were involved, with considerable international participation. The results were published in some 60 volumes in late 1977,[13] the conclusion of the summary volume being: 'The proposed method for the final storage of high level waste glass is therefore deemed to be absolutely safe.' The results of the study were extensively reviewed in Sweden and also in other countries during the first half of 1978. Further work in which specific sites with the required properties were identified resulted in the Swedish Government's acceptance that the conditions of the Stipulation Act had been fulfilled.

Since 1977, there have been many reviews of this subject. The EEC held a Symposium at the end of the first five years of a research programme co-ordinated by the Community in Luxembourg.[14] The subject of the environmental impacts of wastes from all nuclear fuel cycles was also reviewed as part of the International Nuclear Fuel Cycle Evaluation exercise held in 1980, with participation of 66 countries. The INFCE concluded[15] that geological isolation would be satisfactory for all wastes and that the largest, but still low, environmental impact from the thermal reactor fuel cycles would arise from the uranium mine and mill tailings; this would be much reduced if a fast reactor fuel cycle were to be employed, because of the smaller uranium requirements. INFCE pointed out that the safety analyses are limited by the accuracy of the models used to describe the disposal system but that the uncertainty is not such as to affect the general conclusions.

In the UK it was originally proposed to conduct a research and drilling programme aimed at establishing the significant features of a number of rock types. One set of experiments in a granite in Caithness has been completed, but three other applications for planning permission resulted in public inquiries and much public opposition. However, the emerging international consensus that geological isolation would prove entirely satisfactory and the weight of evidence arising from work in many countries led the UK government to stop the drilling programme on the grounds that the option had been proved in principle and that actual operations need not start for many years. Although the conclusion that the feasibility of geological disposal has been proved in principle is undoubtedly correct, further geological field work on possible sites in this country would be required in advance of any decision to proceed to final disposal.

Some general lines of study should continue to improve our knowledge of the behaviour of radioactive waste underground. The first is the improvement of the mathematical model that has been used to predict possible doses to take more realistic account of the processes of liquid diffusion in complicated rock structures, of the effects of the introduction of a source of heat and of the complicated phenomena of absorption of any radionuclides leached out from the glass on rock surfaces. The second is to improve our knowledge of the rates at which the radioactive isotopes would, in practice, be leached from the glass and its near environment. The models used to date have assumed that the glasses would dissolve at the rates measured in flowing water. But the groundwater movement is so slow at considerable depth that the actual values of leaching rates and solubilities of glass in a repository would be determined not so much by the properties of glass but by the maximum solubility in the restricted volumes of water available to interact with it. The effective reduction of leach rate in restricted, as opposed to free-flowing, waters might be a factor of 100 or more, and this is in line with ordinary geological experience. The glasses exhibit solution rates in flowing water similar to those of common rocks such as basalts (Table 4.3),[16] which dissolve slowly in flowing water and undergo weathering reactions on the surface, but which are quite stable at depth where the access of water and oxygen is restricted. Similar effects will limit the reaction rate of glasses. If these predictions are verified by actual measurements of the leaching rates of individual species under the conditions that hold in underground repositories, then it may be expected that the 'engineered barriers' will remain effective for hundreds of thousands of

Table 4.3. *Comparative leaching rates*

Cotswold limestone	34
Glass 189	18
Arthur's Seat basalt	4
Glass 209	2.6
Rockall granite	2
Pitchblende	1
Cornish Aplite Rock	0.4
Malvern Gneiss Rock	0.3

Measurements made in flowing water at 100 °C

years in the absence of acute geological disasters, and one of the criteria for the choice of a suitable site will be the geological stability of the formation.

Although the glass compositions that have been developed therefore promise to give perfectly satisfactory service as a containment medium for the high-level wastes, many other materials have been considered for this duty. One recent idea, put forward by Professor A.E. Ringwood in Australia,[17] is to convert the waste into an intimate mixture of mineral-like materials which can be accepted as durable by analogy with the properties of the natural minerals themselves. Certainly it seems that the synthetic materials that Ringwood has proposed would have lower leach rates than some of the borosilicate glasses. Further work is required to establish how such materials could be made on the ton scale under active conditions. Ceramic materials of this sort may well be developed as second-generation waste forms; the technology of handling high level wastes, like others, will continue to improve, though the present technologies seem adequate.

Disposal under the oceans

The principles of disposal of high-level wastes under the deep oceans would be exactly the same as those already discussed for disposal on land; the engineered barriers would include the containment and any over-cladding used to inhibit corrosion, and the resistance to leaching of the vitrified waste itself. The use of a model of the ocean currents to predict the time at which any leached activity could enter a food chain has been described in the present context of the disposal of low-level waste. The development of such a model to encompass high-level waste disposal on the ocean bed demands increased knowledge both of oceanography and of marine biology. The NRPB[18] has published a preliminary model of the spread of activity from the ocean bed of the wastes that would result from very large world nuclear power programmes. The model used assumed a fairly simple mechanism of dispersion within the ocean and various vehicles of biological concentration were examined as a potential pathway back to man. The maximum radiation doses predicted were those due to the consumption of contaminated marine species and were very small compared with the ICRP limit, particularly if the vitrified waste was contained, as it well could be, in a material that would be corrosion resistant for some centuries. However, the assumptions in this preliminary assessment have to be verified by much further research, particularly into the behaviour of ocean sediments, ocean currents, the phenomenon of upwelling from the bottom and marine food chains. It will take many years for the state of knowledge to advance to the point where higher limits than those presently described by the London Convention might be proposed, and there may in any case be considerable political opposition to any attempt to do so.

The option of emplacement within the deep ocean sediments, below the ocean bed, is intuitively more attractive since these would present an additional barrier between the wastes and man. This concept is similar to that of geological disposal, with the additional factor of dilution in the sea. The ocean sediments are known to be ancient and are thought to have good absorptive properties for many of the active species that might eventually be leached from the vitrified waste by the action of seawater. The US has taken a lead in this research, with international co-operation. A number of areas on the ocean floor have been identified where the sediments are thought to satisfy the desirable criteria, but the work is at a comparatively early stage. Participation in this international programme is led in the UK by the Institute of

Oceanographic Sciences and the Ministry of Agriculture, Fisheries and Food.

Other ideas

Two suggestions have been made to circumvent the need to prove the safety of long-term containment. The first is to reduce significantly the long-lived content of the high-level waste by nuclear incineration of the actinide component, thus converting some long-lived α-emitters into the comparatively short-lived fission products which mainly, though not completely, decay away in 600 years or so. The idea has been examined in an international collaborative programme and some research work continues. Although it might be feasible on nuclear grounds, what would be required would be a very long regime of irradiation in a fast reactor or in another source of neutrons, accompanied by some very difficult chemical separation processes. We have here perhaps an example of an optimisation problem. Long irradiation and complex separation processes would undoubtedly add something to irradiation doses received by the workers who carried them out, and would generate low-level waste streams which would add a little to the additional radiation burden caused by low-level waste disposal. Would this degree of detriment be worth bearing for the advantage of reducing, but not eliminating, the long-term component in waste disposal? Our own present assessments are that on radiological, as well as on economic, grounds there is no balance of advantage in attempting to reduce the actinides in this way, although quite new advances in chemical separation techniques might cause a re-examination.

The second and most revolutionary idea is that the whole policy of the containment and concentration of high-level waste in one or a few places should be reversed, and that the activity should instead be diluted to such an extent that wide dispersal in the environment would be possible. A recent paper[19] by Peuchl develops this idea and shows how the waste, after storage for about 600 years, could be suitable for incorporation into large tonnages of an inert material, such as concrete, and consequently disposed of as low-activity waste. Whether any such scheme would ever be acceptable is of course another matter. Quite apart from pre-scribing actions to be taken several centuries hence, it must be problematical whether spreading sources of radioactivity around the world would be seen as good radiological practice when they could be maintained in a concentrated form until such time as the activity has naturally decayed to low levels! Indeed, the author recognises as much; he proposed a policy of dispersal as a stratagem to get round a political difficulty rather than one leading to any real benefit.

A final perspective

Routes are already available for the safe disposal of low-level waste, and disposal of the high-level and intermediate-level wastes can now be seen as a tractable problem. The volumes are low and the materials can be converted into stable and durable forms by the application of familiar tech-nologies. The levels of activity remain high for, at most, some hundreds of years and complete containment should cer-tainly be possible for that period of time. The amount of radioactivity that remains after some centuries is a minute fraction of that which occurs naturally near the earth's sur-face. The knowledge of containment conditions that we already have enables us to state with confidence that we shall be able to design and build a repository or a series of repositories in geological formations on land when we need to do so with a high degree of safety, and that there are good prospects that the ocean disposal options will be found to be equally acceptable. No human disaster can be foreseen. The criterion of safety that has been implicit in the studies summarised in this chapter is that no *individual* in the distant future will receive more than the current limit of annual radiation dose for members of the public, which is set by the ICRP. Remembering that this corresponds to less than $\frac{1}{10\,000}$ chance of dying of cancer for each year of exposure, and that the application of such limits to individual doses has resulted in *average* doses to populations that are more than a hundred times lower than this, the risks are seen to be minute. It is an extraordinary standard to be even contemplating; the economic consequences would be serious if similar standards were to be applied to other industries or activities.

Looking to the problems of environmental control that will beset mankind in the future, the radioactive waste problem is a limited and comparatively minor one. Because the volumes of waste are small, it is a task which can be mastered by a suitable concentration of technology on a narrow front in a limited number of locations. The numbers of people engaged in the task will be small. Technologists can always do that sort of thing when there is no fundamental scientific difficulty; after all, a suitable concentration of resources brought men back alive from the moon and brought the Space Shuttle back to earth. The really difficult problems facing our descendants will be those involving huge numbers of people – over-population, famine, de-forestation, decreasing natural resources; the rising level of carbon dioxide, if it ever looked dangerous, would be difficult to reverse. All these will prove as intractable as the problem of chronic poverty in the Third World is today. To solve problems with such wide dimensions, technologists may carry heavy responsibilities, but technology alone will be insufficient. You cannot solve them by a concentration of effort on a narrow front. But technologists can properly claim that they can deal safely with radioactive wastes, and the organisations and regulatory bodies that exist are perfectly adequate to ensure that they will do so.

5
Looking to the future

This book has been concerned with the responsibilities of technologists to avoid undue harm arising from the development of new products and processes. This may be seen as a rather limited view; the most obvious responsibilities that technologists must shoulder in the face of a growing world population and the realisation that the earth's resources are finite is to contribute to the wise use of those resources by measures of conservation and of substitution, so as to preserve and develop the world's economy. The question of future energy supplies was considered briefly in the first chapter. It is a vital and fundamental question, since all plans for the use of lower-grade ores, for the reclamation of land, or for the re-cycling of materials depends on the availability of appropriate sources of energy at prices that allow for economic use. The case for nuclear power is that it is a proven technology that allows us to preserve some of our limited fossil fuel resources for the benefit of future generations without incurring any economic penalty in doing so; from the point of view of satisfying a range of human needs, oil and coal are flexible resources, while uranium ore is of limited use.

Whenever new technologies are introduced into the modern industrialised world, questions of safety and of environmental consequence will be raised. In the previous chapters, the evolution of ideas on how to deal with three general problems has been traced – the need to protect people from harmful concentrations of toxic agents; the reduction of the risks of accidents to acceptable levels; and the need to protect the environment against future deterioration. In the context of the nuclear industry, scientists and technologists have had to evolve satisfactory standards for exposure to radiation; to develop the methodology of assessing the risks of accidents to complex plant; and to deal with the arising wastes, which, typically, are radioactive. In considering all of these, three separate but linked groups of questions can be identified.

First there are the scientific questions – the easiest in many ways, complex though they are. The susceptibility of the population to very low doses of agents known to be harmful in large concentrations; the development of a data base and of methods of calculating the probabilities and consequences of accidents that are very unlikely events; the development of models of migration through the earth and the sea that allow for extrapolation from the present into the future – all such subjects, which apply to many technologies, can only be pursued through long and patient study by experts in many sciences who are intent on reaching a consensus. This is the normal process of scientific advance, conducted by the learned societies and international associations. In the nature of things, 'final' answers cannot be expected, but it is necessary for scientists and engineers engaged in these fields to hold to the proper disciplines of scientific debate and, acknowledging that the understanding of complex phenomena is always incomplete, to avoid exaggeration and a lack of perspective in reporting new results. As Sir Alan Cottrell has said in his recent book, a breach of this discipline is tantamount to misleading the public, just as is any attempt to keep silent about matters of proper public concern.[1]

Secondly, there is the stage of professional assessment of the scientific base, with a view to arriving at a practical set of rules and regulations. So far as radiation exposure is concerned, this is the task undertaken by the International Commission and, in the UK, by the National Radiological Protection Board. The safety of plant is the responsibility of the operator, but the standards to be met must satisfy the Nuclear Installations Inspectorate of the Health and Safety Executive. Waste management practices are controlled by those government departments responsible for environmental matters. Inevitably, assessments of safety standards involve value judgements of what constitutes an acceptable level of risk.

Thirdly, decisions on the level of safety deemed acceptable to society and on what money or resources should be allocated to attain it must be the responsibility of government. However, there is also a need to ensure that proper public discussion takes place before political decisions can be taken. In a democracy, decisions affecting the safety of the public must be taken openly and by bodies which are accountable to the public. How to achieve a proper sense of public participation is clearly difficult. One device that has been used is the appointment of bodies of people, independent both of the industry and of the regulatory authorities, to advise the relevant Minister directly. This is the function of the Radioactive Waste Management Advisory Committee and of the Advisory Committee on the Safety of Nuclear Installations. Such bodies cannot duplicate the detailed work of the professionals in the departments of the industry concerned with safety or in the Radiochemical Inspectorate of the Department of the Environment, the National Radiological Protection Board or the Nuclear Installations Inspectorate, but they can question and expose the philosophy on which their work is based, or form an opinion on the public's reaction to their conclusions. A similar function can be carried out, of course, by the Select Committees set up by Parliament.

Some more direct method of involvement of interest groups and members of the public is necessary, particularly in matters affecting local populations. The method that has been used in the UK is to hold a planning enquiry in public on a specific local proposal. This has the advantage of allowing the structured development of an argument in a confrontational mode, with cross-examination of witnesses and participation of members of the public. The Windscale Enquiry on the proposal by BNFL to site a new reprocessing plant at Sellafield and the present Enquiry on the CEGB's plan to build a pressurised water reactor at Sizewell are major examples which affect the future of the nuclear industry in the UK. It must be questioned how far such enquiries are fitted to deal both with national and local issues, though these are not easily separated – as, indeed, is the case in the siting of coal pits, airports or motorways. Some limits, however, must be set. At one extreme, a criticism of Judge Parker's Report on the Windscale Enquiry was that the objectors and the proponents were not arguing from the same basic assumptions on the nature and values of our society. Surely no enquiry of this type can be expected to debate such a large issue; the current social consensus must be assumed as a basis for planning. At the other extreme, a planning enquiry is not the right context in which to debate the scientific basis on which regulations and safety cases are based, though full explanations of the reasons behind regulations must be available. Those who wish to question the assessment of the fundamental scientific results should be enabled to do so through the mechanisms of the established learned societies, and these organisations – well used as they are to vigorous debate – should be able to ensure that all views are fairly weighed.

What is important is that the three types of question – scientific, professional assessment and socio–political – be addressed in appropriate institutions if progress is to be made. Furthermore, the general questions of the criteria to be used and the allocation of resources to public health and safety in different fields should be widely debated and better understood than they are today. Any such debate about safety standards must be concerned with the costs of safety measures and with the marginal cost of attaining higher standards, in different contexts, if we are to derive the most benefit from our technological opportunities and apply our resources in an efficient and a humane way. It is an inescapable responsibility of technologists to supply the information on which such judgements can be made in as full, objective and clear manner as possible. I hope this book has illustrated that technologists in the nuclear industry are attempting to meet that obligation, and that the safety standards to which the industry is working are high and are based on a great deal of information and experience.

Annex

Fig. A1.1

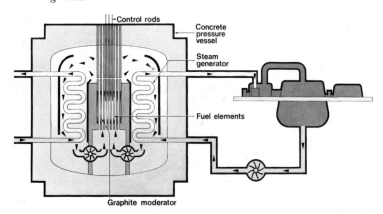

Advanced gas-cooled reactor (AGR)

Fig. A1.2

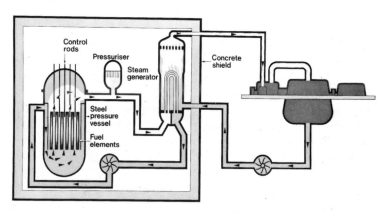

Pressurised water reactor (PWR)

Fig. A1.3

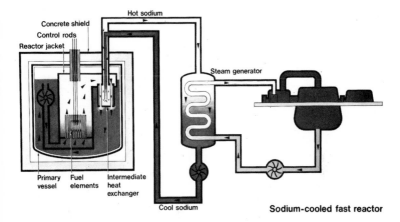

Sodium-cooled fast reactor

Glossary

ACRS: Advisory Committee on Reactor Safeguards, USA.

ALARA: the need to keep radiation exposures *As Low As Reasonably Achievable* – one of the ICRP principles of radiological protection.

actinides: a series of heavy chemical elements, all of which are radioactive. It includes the elements uranium and plutonium.

alpha particles (α): see radiation.

BEIR: Committee on the Biological Effects of Ionising Radiation, National Academy of Sciences, USA.

beta particles (β): see radiation.

BNFL: British Nuclear Fuels Limited.

carcinogenesis: the production and development of cancer.

CEGB: Central Electricity Generating Board (England and Wales). The other generating boards in the UK are the South of Scotland Electricity Board, the North of Scotland Hydro-Electric Board and the Northern Ireland Electricity Service.

chain reaction: a reaction whose products set off further reactions of the same kind. In a fission chain reaction, neutrons are produced which are able to initiate further fission reactions in fissile nuclei.

cladding: the can in which nuclear fuel is sealed to protect it and retain its fission products.

collective dose: see radiation dose.

control rods: rods of a neutron-absorbing substance which are used to control the reactivity of a reactor. The degree to which the control rods are inserted into the core of the reactor controls the number of free neutrons available and hence the rate at which the chain reaction proceeds.

critical group: a term used in radiological protection for those members of the public likely to be exposed to the largest dose of radiation as a result of any particular operation.

decay heat: heat produced by the radioactive disintegration of fission products, which continues even after the chain reaction has stopped.

130

design basis event: the maximum expected event (e.g. maximum load) that a structure is designed to resist with an adequate safety margin.

dose: see radiation dose.

EPRI: Electrical Power Research Institute, USA.

ERL: Emergency Reference Level: the level of radiation dose at which public authorities are required to consider protective measures in the event of an accident.

event tree analysis: the analysis of the possible events that may follow a single initiating event in a complex plant.

fast reactor: a reactor in which the chain reaction is sustained by fast neutrons. In fast breeder reactors, new fuel is bred by the absorption of surplus neutrons in the non-fissile isotope U^{238}. (See Fig. A1.3)

fission: the spontaneous or induced disintegration of the nucleus of a heavy atom into two or more lighter ones (the fission products), with the release of energy.

fuel reprocessing: the process whereby used nuclear fuel is separated into uranium, plutonium and radioactive waste products.

fusion: the uniting of two light nuclei into one heavier one, with the release of energy.

gamma radiation (γ): see radiation.

Gigawatt: a unit of power, equal to one thousand million watts. It could supply one million one-kilowatt domestic electric heaters.

hereditary effect (*also, genetic effect*): an effect capable of being passed on from one generation to another; in radiation protection a possible result of radiation damage to the testes or ova.

IAEA: International Atomic Energy Agency, based in Vienna, consisting of 111 countries.

ICRP: International Commission on Radiological Protection.

INFCE: International Fuel Cycle Evaluation, an international study which reported in 1980.

ionisation: see radiation.

isotopes: nuclei of a given chemical element with different atomic masses, resulting from the presence of different numbers of neutrons.

leaching: the removal of material by percolating water.

London Convention: the Convention on the Prevention of Marine Pollution by Dumping of Wastes, signed since 1972 by 47 countries.

mutation: a change in the characteristics of an organism produced by an alteration of the hereditary material.

National Registry of Radiation Workers: a registry set up by NRPB in 1976 to collect data on radiation doses and mortality of radiation workers in the UK.

NEA/OECD: Nuclear Energy Agency of the Organisation of Economic Co-operation and Development.

neutrinos: sub-atomic particles emitted during some radioactive decays. Neutrinos interact very weakly with matter.

neutrons: see radiation.

NII: Nuclear Installations Inspectorate, part of the Health and Safety Executive, responsible for the licensing of UK nuclear plant.

NPT: Non-Proliferation Treaty.

NRC: Nuclear Regulatory Commission, USA.

NRPB: National Radiological Protection Board, UK.

plutonium: a fissile element produced by transmutation of U-238 after absorbing a neutron. Used in fast reactor fuel.

probabilistic risk assessment: a mathematical technique used to determine the likelihood of faults occurring and their possible outcome.

radiation: the term used in this book for ionising radiation, that is, particulate or electromagnetic radiation that produces ionisation (the removal of one or more electrons from an atom, leaving positively charged ions) in material through which it passes. The ionising radiations are:

  alpha particles: the nuclei of atoms of the element helium;

  beta particles: electrons;

  gamma radiation and X-rays: electromagnetic radiations similar to light and radio waves;

  neutrons: neutral particles present in all nuclei except hydrogen.

radiation background: the natural radiation to which people are exposed.

radiation dose: the quantity of radiation absorbed by an individual or by a specific organ. Collective dose is a measure of the radiation dose received by a population; it is the sum of the doses received by all the individuals in that population.

radiation dose equivalent: a measure of the biological effectiveness of a dose of radiation.

radiation units: the units used in this book are the rad and the rem. The rad is the unit for the quantity of radiation absorbed per unit weight of absorbing tissue and is $1/100$ joule per kilogram. The rem is the unit for dose equivalent (qv) and is also $1/100$ joule per kilogram. It is the product of the absorbed dose and a quality factor which expresses the different effectiveness of the different types of radiation in producing specific biological effects. New units of dose and dose equivalent, the gray and the Sievert, are coming into use.

radioactivity: the transformation of unstable nuclei, resulting in the emission of particulate or gamma radiation.

radiological protection: the protection of individuals from the effects of ionising radiation.

RWMAC: Radioactive Waste Management Advisory Committee, UK.

somatic: relating to the body of an animal as distinct from the germ cells. A somatic effect of radiation is one affecting the individual receiving the radiation, as distinct from a hereditary effect, which affects subsequent generations.

source term: the quantity of radioactive material released in an accident.

specific activity: the radioactivity per unit mass or volume.

stochastic: governed by statistics. The stochastic effects of radiation are those in which the dose governs the probability of the effect occurring, not the severity of the effect itself. The non-stochastic effects are those in which the dose governs the severity of the effect.

thermal reactor: a reactor in which the chain reaction is sustained by thermal neutrons, neutrons which have been slowed down by passage through a moderator. The principal thermal reactor types are illustrated in the Annex (see Figures A1.1, A1.2)

UNSCEAR: United Nations Scientific Committee on the Effects of Atomic Radiation.

UKAEA: United Kingdom Atomic Energy Authority.

uranium: the heaviest naturally occurring element, consisting mainly of U^{238} together with 0.7% of the fissile isotope U^{235}. Enriched uranium is uranium in which the proportion of U^{235} has been increased.

vitrification: the process of conversion to a glassy solid.

References

Chapter 1

[1] Gerholm, T.R., 'Long-range energy demand problems and perspectives', in *Proc. 11th World Energy Conference*, Munich, vol. RTB, pp. 251–90, September 1980.

[2] Smil, V. & W.E. Knowland (eds.), *Energy in the Developing World: The Real Energy Crisis*, Oxford University Press, 1980.

[3] *The World in Figures, Economist*, London, 2nd edn, 1978.

[4] Landsberg, H. (ed.), *Energy: The Next Twenty Years*, Ballinger Publishing Co., Cambridge, USA, 1979.

[5] Posner, E., 'Reception of Röntgen's discovery in Britain and USA', *British Medical Journal*, 4, 357–60, 1970.

[6] Eve, A.S., *Rutherford*, Cambridge University Press, p. 93, 1939.

[7] *Sources and Effects of Ionizing Radiation.* Report of the United Nations Scientific Committee on the Effects of Atomic Radiation (UNSCEAR), United Nations, New York, 1977.

[8] *The Effects on Populations of Exposure to Low Levels of Ionizing Radiation: 1980.* Report of the National Research Council Committee on the Biological Effects of Ionizing Radiation (BEIR), National Academy Press, Washington DC, 1980.

[9] 'Recommendations of the International Commission on Radiological Protection', (ICRP Publication 26) *Annals of the ICRP*, 1 (3), 1–53, 1977.

[10] Loewe, W.E. & E. Mendelsohn, 'Revised dose estimates at Hiroshima and Nagasaki', *Health Physics*, 41 (4), 663–6, 1981.

[11] Frigerio, N.A. *et al*, 'Carcinogenic hazard from low-level, low-rate radiation', Argonne Radiological Impact Programme (ARIP) Part 1, *ANL/ES-26 (Pt. 1)*, Argonne Naval Lab., Ill., USA, 1973.

[12] Rose, K.S.B., 'Review of health studies at Kerala', *Nuclear Energy*, 21 (6), 399–408, 1982.

[13] High Background Radiation Research Group, China, 'Health survey in high background radiation areas in China', *Science*, 209 (4451), 877–80, 22 August 1980.

[14] *Federal Research on the Biological and Health Effects of Ionizing*

Radiation. Report of the National Research Council Committee on Federal Research on the Biological and Health Effects of Ionizing Radiation, National Academy Press, Washington DC, 1981.

[15] Handler, P., Dedication Address, North Western University Cancer Centre, 18 May 1979.

Chapter 2

[1] *Journal of the Roentgen Society*, p. 113, October 1915.

[2] Chapter 1, Ref. 9.

[3] Office of Population Censuses and Surveys, *Mortality Statistics, Accident and Violence Series DH4*, HMSO.

[4] Royal College of Physicians, *Smoking or Health, Third Report*, Pitman Medical, London, 1977.

[5] Pochin, E.E., 'The acceptance of risk', *British Medical Bulletin*, 31 (3), 184–90, 1975.

[6] Taylor, F.E. & G.A.M. Webb, 'Radiation exposure of the UK population', *NRPB-R77*, National Radiological Protection Board, HMSO, 1978.

[7] *Ionizing Radiation: Sources and Biological Effects.* 1982 Report of the United Nations Scientific Committee on the Effects of Atomic Radiation (UNSCEAR 1982), United Nations, New York, 1982.

[8] Ministry of Agriculture Fisheries and Food, *Radioactivity in Surface and Coastal Waters of the British Isles, 1981*, Lowestoft, 1983.

[9] *Living with Radiation*, National Radiological Protection Board, Chilton, Oxon, 2nd edn, 1981.

[10] Darby, S.C. *et al.*, 'The genetically significant dose from diagnostic radiology in Great Britain in 1977', *NRPB-R106*, National Radiological Protection Board, HMSO, 1980.

[11] Pochin, E.E., 'The need to estimate risks', *Physics in Medicine and Biology*, 25, 1–12, 1980.

[12] Cohen, A.V. & D.K. Pritchard, 'Comparative risks of electricity production: a critical survey of the literature', *HSE Research Paper 11*, HMSO, London, 1980.

[13] Webb, G.A.M. & A.S. McLean, 'Insignificant levels of dose: a practical suggestion for decision making', *NRPB-R62*, National Radiological Protection Board, HMSO, 1977.

[14] 'Cost-benefit analysis in optimising the radiological protection of the public: a provisional framework', *ASP4 (1981)*, National Radiological Protection Board, HMSO, 1981.

[15] Chapter 1, Ref. 15.

Chapter 3

[1]   Royal Society, *Risk Assessment. A Study Group Report*, The Royal Society, London, 1983.

[2]   Hinton, C., Axel Axson Johnson Lecture, Stockholm, 15 March 1957.

[3]   Farmer, F.R., 'Reactor safety and siting: a proposed risk criterion', *Nuclear Safety*, 8 (6), 539–48, 1967.

[4]   'Reactor safety study: an assessment of accident risk in US commercial nuclear power plants', *WASH-1400*, ('Rasmussen Report'), Nuclear Regulatory Commission, Washington DC, 1975.

[5]   Lewis, H.W. *et al.*, 'Risk Assessment Review Group Report to the US Nuclear Regulatory Commission', *NUREG-CR-0400*, Nuclear Regulatory Commission, Washington DC, 1978.

[6]   Leverenz, F.L. Jr. & R.C. Erdmann, 'Comparison of the EPRI and Lewis Committee review of the reactor safety study', *EPRI-NP-1130*, Electric Power Research Institute, Palo Alto, Cal., 1979.

[7]   *Risk Study on Nuclear Power Stations.* A study made for the Federal German Ministry of Research and Technology by the Gesellschaft für Reacktorsicherheit (in German), Verlag, Cologne, 1980. English translation available as 'German risk study – main report. A study of the risk due to accidents in nuclear power plants' *EPRI NP-1804-SR*, Electric Power Research Institute, Palo Alto, Cal., 1981.

[8]   'Zion probabilistic safety study', Commonwealth Edison Company, Illinois, USA, 1981.

[9]   Kelly, G.N. & R.H. Clarke, 'An assessment of the radiological consequences of releases from degraded core accidents for the Sizewell PWR', *NRPB-R137*, National Radiological Protection Board, HMSO, 1982.

[10] Marshall, W., 'Talking about accidents', *Atom*, No. 312, pp. 210–15, October 1982.

[11] Kemeny, J.G. (Chairman), *Report of the President's Commission on the Accident at Three Mile Island*, Washington DC, October 1979.

[12] Locke, J.H. *et al.*, *Canvey: An Investigation of Potential Hazards from Operations in the Canvey Island/Thurrock Area*, Health and Safety Executive, HMSO, 1978.

[13] Safety and Reliability Directorate (United Kingdom Atomic Energy Authority) *Canvey: A Second Report: A Review of Potential Hazards from Operations in the Canvey Island/Thurrock Area Three Years after Publication of the Canvey Report*, Health and Safety Executive, HMSO, 1981.

[14] Fryer, L.S. & R.F. Griffiths, 'Worldwide data on the incidence of multiple-fatality accidents', *SRD-R-149*, Safety and Reliability Directorate (United Kingdom Atomic Energy Authority), HMSO, 1979.

[15] Crick, M.J. & G.S. Linsley, 'An assessment of the radiological impact of the Windscale reactor fire, October 1957', *NRPB-R135, and 135 (addendum)*, National Radiological Protection Board, HMSO, 1982.

[16] 'Toward a safety goal: discussion of preliminary policy considerations', *NUREG-0764*, Nuclear Regulatory Commission, Washington DC, 1981.

[17] Nuclear Installations Inspectorate, *Safety Assessment Principles for Nuclear Power Reactors*, Health and Safety Executive, London, 1979.

[18] 'Design safety criteria for CEGB nuclear power stations', *CEGB/HS/R-167/81 (revised)*, Nuclear Health and Safety Dept., Central Electricity Generating Board, London, 1981.

[19] Dunster, H.J., 'The value and limitation of quantitative safety goals'. Paper presented at the American Nuclear Society Meeting on Probabilistic Risk Assessment, Port Chester (New York), USA, 20–23 September 1981.

[20] Slovic, P. *et al.*, 'Perceived risk: psychological factors and social implications', *Proceedings of the Royal Society, London*, A375, 17–34, 1981.

[21] Thomas, K., 'Comparative risk perception: how the public perceives the risks and benefits of energy systems', *Proceedings of the Royal Society, London*, A376, 35–50, 1981.

[22] Cohen, B.L., 'Society's valuation of life saving in radiation protection and other contexts', *Health Physics*, 38 (1), 33–51, 1983.

[23] Kletz, T.A., 'Benefits and risks: their assessment in relation to human needs', *Endeavour, New Series*, 4, 46–51, 1980.

[24] Siddall, E., 'Control of spending on nuclear safety', *Nuclear Safety*, 21, 451–60, 1980.

[25] O.Donnell, E.P. & J.J. Mauro, 'A cost–benefit comparison of nuclear and nonnuclear health and safety protective measures and regulations', *Nuclear Safety*, 20, 525–40, 1979.

[26] Ashby, E. Lord, *Reconciling Man with the Environment*, Oxford University Press, 1977.

Chapter 4

[1] 'The control of radioactive wastes', *Command Paper 884*, HMSO, London, November 1959.

[2] *A Review of CMND 884: 'The Control of Radioactive Wastes'*. A report by an expert group made to the Waste Management Committee, Department of the Environment, London, September 1979.

[3] 'Radioactive waste management', *Command Paper 8607*, HMSO, London, July 1982.

[4] Duncan, A.G. & S.R.A. Brown, 'Quantities of waste and a strategy for treatment and disposal', *Nuclear Energy*, 21 (3), 161–6, 1982.

[5] Beale, H. *et al.*, 'The disposal of low and intermediate level wastes in the UK'. Paper presented at the International Fuel Cycle Conference, Atomic Industrial Forum, Geneva, 31 May–1 June 1983 (to be published).

[6] Lewis, J.B., 'The case for deep sea disposal', *Atom*, No. 317, 49–52, March 1983.

[7] Radioactive Waste Management Advisory Committee, *Fourth Annual Report*, HMSO, London, 1983.

[8] Royal Commission on Environmental Pollution (Chairman Sir B. Flowers), 'Nuclear power and the environment. Sixth report', *Command Paper 6618*, HMSO, September 1976.

[9] Ewart, F.T., 'Fuel inventories and derived parameters for the reactor systems CDFR, LWR, AGR and Magnox', *AERE-R-10037*, United Kingdom Atomic Energy Research Establishment, HMSO, 1981.

[10] Hill, M.D. & P.D. Grimwood, 'Preliminary assessment of the radiological protection aspects of disposal of high level radioactive waste in geological formations', *NRPB-R69*, National Radiological Protection Board, HMSO, 1978.

[11] Hill, M.D., 'Analysis of the effect of variations in parameter values on the predicted radiological consequences of geologic disposal of high-level waste', *NRPB-R86*, National Radiological Protection Board, HMSO, 1979.

[12] Hill, M.D. & G. Lawson, 'An assessment of the radiological consequences of disposal of high level waste in coastal geologic formations', *NRPB-R108*, National Radiological Protection Board, HMSO, 1980.

[13] Kärn-Bränsle-Säkerhet, *Handling of Spent Nuclear Fuel and Final Storage of Vitrified High Level Reprocessing Waste*, 5 vols., A.B. Telepan, Sweden, (1978).

[14] Simon, R. & S. Orlowski (eds.), *Radioactive Waste Management and Disposal*. Proceedings of the First European Community Conference, Luxembourg, (EUR-6871), Harwood Academic, Chur, 1980.

[15] International Fuel Cycle Evaluation, *Waste Management and Disposal.* Report of INFCE Working Group 7, International Atomic Energy Agency, Vienna, 1980.

[16] Marples, J.A.C. *et al.*, 'The leaching of solidified high level waste under various conditions', *Proc. 1st Conference on Radioactive Waste Management and Disposal*, Luxembourg, pp. 307–23, 1980.

[17] Ringwood, A.E. *et al.*, 'Immobilisation of high level nuclear reactor wastes in SYNROC', *Nature*, 278 (5701), 219–23, 15 March 1979.

[18] Grimwood, P.D. & G.A.M. Webb, 'Assessment of the radiological protection aspects of disposal of high level waste on the ocean floor', *NRPB-R48*, National Radiological Protection Board, HMSO, 1976.

[19] Peuchl, K.H., 'The nuclear waste problem in perspective', *Nuclear Engineering International*, 20, 950–4, 1975.

Chapter 5

[1] Cottrell, A. Sir, *How Safe is Nuclear Energy?*, Heinemann, London, 1981.

Index